150
Most-Asked
Questions About
Menopause

150
Most-Asked
Questions About
Menopause

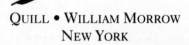

WHAT WOMEN
REALLY WANT TO KNOW

Ruth S. Jacobowitz

QUILL • WILLIAM MORROW
NEW YORK

Copyright © 1993 by Ruth S. Jacobowitz

Library of Congress Cataloging-in-Publication Data

Jacobowitz, Ruth S.
150 most-asked questions about menopause : what women really want to know / Ruth S. Jacobowitz.
p.   cm.
Includes index.
ISBN 0-688-14768-2
1. Menopause—Miscellanea. I. Title. II. Title: One hundred-fifty most-asked questions about menopause.
RG186.J32   1992
612.6'65—dc20   92-16608
                                                   CIP

Printed in the United States of America

3 4 5 6 7 8 9 10

BOOK DESIGN BY JAYE ZIMET

*For my mother, Claire Scherr, who exemplifies
having pizzazz and a zest for life*

Acknowledgments

This book was written in honor of and at the request of the women I was lucky enough to meet as I lectured around the country on women's midlife health issues. I thank each and every woman who came to the lectures and who took the time to share their personal stories with me. Sponsored by many prestigious medical organizations, these free women's health programs were made possible through their continuing commitment to consumer education. I am deeply appreciative of that sponsorship and support. I am also grateful to the physicians who generously shared the podium with me: Mary K. Beard, M.D., Sarah Berga, M.D., Peter Hickox, M.D., Jon S. Nielsen, M.D. Roger D. Matthews, M.D., William D. Schlaff, M.D., Neil Wolfson, M.D., and Wulf H. Utian, M.D. This book is for all of them!

My sincere special thanks also to the world authority on menopause, Dr. Wulf H. Utian, my coauthor on *Managing Your Menopause*, and my friend. He always believed that we women have a right to good health care and, in a way, he started me off on my mission.

Nothing is created in a vacuum. So many others participated in making the programs a success and making this book a reality. I particularly want to thank Jeff, Pam, Warren, Lisa, Laura, Betsy, Joan, Bobbie, Marge, and all the others who helped me and who graciously shared information with me. My gratitude goes to my editor, Toni Sciarra. Toni is wise beyond her years, talented beyond her experiences, and nurturing beyond the capacity of most people I know. My deep appreciation also goes to my assistant and secretary, Marilyn Morgan, who shared my excitement in creating this book. I have been fortunate in all these relationships.

I was raised in a female world. I have two sisters, Harriet Lewis and Susan Rosenberg. I have three daughters and one granddaughter. But it was my terrific husband, sons-in-law, and grandsons who taught me the real differences in behaviors between males and females. I apologize to them all for all the family time that became lecturing or writing time. Thanks to my kids: Jan Jacobowitz, Alvin, and Jeffrey Asher Lodish; Jody, David, Claire Michelle, and Jake Cremer Austin; and Julie, Lowell, and Michael Aaron Potiker. You're quite simply the greatest support group any mother could have! And a special note to Lowell—you made me computer literate in the first place and you so kindly and gently continue to unscramble my technological goofs—I could not have finished this book on time without you.

Acknowledgments

Last, and first, and always, to Paul, my husband of more than forty years, my undying gratitude. You not only survived my menopause, but you continue to support me in all my endeavors on behalf of other women as well.

Ruth S. Jacobowitz

Contents

Contents

Throughout this book references are made to estrogen replacement therapy (ERT) and to hormone replacement therapy (HRT). Usually ERT refers to the use of estrogen alone. In combination with progestin, I refer to it as HRT.

HRT is most often prescribed for women with an intact uterus to safeguard them against endometrial cancer, cycling the hormones in much the same physiologic manner as their own bodies had done when their ovaries were producing these hormones naturally. Using the hormones in this way is currently thought to protect the endometrium (lining of the uterus) from hyperplasia and cancer by assuring that it is shed on a regular basis.

NOTE: A 1992 background paper, "The Menopause, Hormone Therapy, and Women's Health," published by the Congress of the United States Office of Technology Assessment, has changed some of the nomenclature regarding estrogen replacement therapy because "some consumer groups oppose the notion that the menopause causes an estrogen deficiency that requires replacement." Throughout their report, they use ET (rather than ERT) to indicate estrogen therapy, and CHT (rather than HRT) to indicate combined hormone therapy. It is possible that this will become the nomenclature of the future. While I am supportive of the consumer view, these therapies are referred to in this book as ERT and HRT, since as of this writing, women are most likely to hear them referred to in this way by medical professionals.

Before You Begin This Book

I ran headlong into menopause. No thought and no preparation. That's because I pretended not to notice the subtle changes that had been occurring in me over the previous few years. Hindsight, as they say, is 20/20. When I finally looked back, I realized that I had subconsciously chosen not to label those changes as symptoms or to consider myself premenopausal.

Although, as a medical journalist and former vice-president of a large teaching hospital I knew many of the symptoms of impending menopause, I chose to relate none of them to me. When the odd sweats and the nervousness began to alter significantly the otherwise excellent quality of my life, I readily blamed them on something else—anything else. Professional burnout, stress, and empty-nest syndrome were popular labels in the mid-1980s, and they were acceptable to me. I used them freely.

A hysterectomy ten years earlier had relieved me of my uterus and with it my periods, so there were no menstrual changes to challenge my illusions of undiminished youth. There I was, enjoying the prime of everything in my life—the peak of my career and the most comfortable and rewarding years of my marriage and of motherhood. Our three daughters were guiding themselves on their self-prescribed courses through the channels of education, marriage, and motherhood. Life had never been better! How long could this utopia continue? Just until I got my first hot flash.

Having written for newspapers and magazines early in my career, I continued to think of a hot flash as a news tip. Not so, I learned, as my first flash was followed by an army of others, some so profound as to wake me from my sound sleep in moist discomfort. Sleepless nights were followed by listless days and mood swings that, like a pendulum, marked the tempo of these unwelcome changes in my life.

Changes in my life—aha, that's it!—I thought. This was the "change" that my mother and her friends had whispered about. I'm going through The Change. That idea was fraught with negative images—images of wrinkled, stooped, nervous women, who were quietly "put to pasture," or grandmothers who had to wear protective pads so that they could hug their grandchildren without leaking urine. Those negative images had been gathered from friends, family members, business associates, and from every form of the media, throughout my youth and my middle years. I fought those images with all the ammunition I had. I didn't want my life to change unless it was to change for the better. When symptom piled upon symptom, however, I could no longer hold my pose of denial. Eventually, I literally "lost it" and learned the hard way that I had to come to grips with and take control of what was happening to me.

I woke up in the middle of the night back in 1985, fearing that I was going crazy. Trembling and perspiring, shaking on the inside from palpitations and on the outside from the chills that alternated with the intense heat I felt, I was scared out of my wits. For the first two weeks that I was in that condition, I hid out at home. Finally, when my husband and children could bear the state I was in no longer, I looked for a doctor who was new in the community and not affiliated with the hospital in which I worked. After all, if you think you are losing your mind, you don't want to share it with your friends or business associates. The new young physician treated me for about a month for everything *but* menopausal symptoms. Finally, I got help from a menopause specialist, and in a couple of weeks I was my old self again.

That's what this book is about. It's about getting the information you need about the "M" word, so that your life, when it gets ready for what has euphemistically been called the "change," is going to change for the better!

My lack of knowledge and preparedness drove me

through eight weeks of hell followed by six months of fear. How could all this bad stuff be happening to me just as I was getting to the "good part" of life?

Ultimately, the good that this ill wind brought was the realization that I should take my life experience as a writer and as a hospital public affairs vice-president and redirect it. I knew quite clearly that I needed to become an activist on behalf of women, encouraging them to become informed about the issues surrounding menopause.

What followed was the publication of *Managing Your Menopause*, the book which I coauthored with my physician, Dr. Wulf H. Utian, a world-renowned expert in the care of menopausal women. In that book, I detailed my experience and, with his cutting-edge knowledge of the subject, we set forth everything that *we* believed women needed to know in order to manage their menopause.

The results of that book surprised me. I was sought out as a speaker on the subject of menopause and was interviewed throughout the country. I came to realize that many women wanted to hear about menopause from a woman—especially from a woman who had been through it, one who had had a difficult time, and who had come out perfectly fine postmenopausally. What I learned from the consumer education programs at which I spoke and from interviews and conversations with women was *what women really want to know* about menopause.

In *150 Most-Asked Questions About Menopause,* I have tried to provide information in the way women told me they want to receive it. Included in this book are some of the most frequently asked questions from the more than ten thousand women to whom I spoke at educational programs held throughout the United States from 1990 through 1992. The programs from which these questions and answers come all follow similar formats. Guests are invited to the programs through the media or by flyers sent to them from the sponsoring medical institutions or organizations. Most of the programs are held in the evening for the maximum convenience of women who work, and they are very crowded.

Also included in this book is pertinent information from a recent Gallup study of women's and men's attitudes, knowledge, and experience with menopause, as well as material from a 1991 National Family Opinion Survey that tells us what women say they know or need to know about postmenopausal osteoporosis. You may be surprised at how candidly the subject of menopause is being discussed, as well as by what some of this newly acquired information reveals about the differing perceptions of men and women regarding menopause.

I've chosen a question-and-answer format for this book because that's how women, who are busier and working harder today than ever before, tell me they want to learn the answers—in short "bites" located easily and quickly. Women want to understand menopause clearly and then be able to review the information on an as-needed basis. They also want sound information to serve as the basis for turning their time with their physicians into "quality time."

Menopause affects each woman differently. Some women (about 15 percent) make this rite of passage easily, moving comfortably and gracefully from the premenopausal to the postmenopausal years. Other women (another 15 percent), like me, have a rough time and need to sort out what may be happening to them. Then there is the group in the middle—approximately 70 percent—who, as one good female friend says, "don't feel like they have a sickness, but they don't feel like they have a wellness, either." They need information, too.

For all women, knowledge is power. And it's never too soon to learn about life's rites of passage. Changes for some of us can begin in our thirties, and for others menopause may come even earlier as the result of a hysterectomy and bilateral oophorectomy. There are helpful answers in this book for women of every age: for young women, for women in the middle years, and for those of us who are enjoying what Margaret Mead called postmenopausal zest. There are so many women who need to know about this time of life!

By the turn of the century, the number of women in the menopausal age range (forty-five to fifty-five) will swell to fifty million. Some statistics tell us that by the year 2000, one third (other studies put that number closer to one half) of the female adult population in the United

States will be in that age range. Right now, three thousand five hundred women per day are pouring into the menopausal category. The good news is that this is the first generation of women who will enter menopause with all the practical information they need available to them, so that they can achieve a first-rate second half of life. That's so important, because new statistics show that these women still have a full one third of their lives remaining to be lived *after* menopause. Maximizing the quality of those years is what *150 Most-Asked Questions About Menopause* is all about.

I've tried to group the questions by subject matter. That wasn't always easy because the questions women ask, like menopause itself, often overlap into many areas. The questions are presented just as the women asked them at the programs, and they are answered with the information that the physician panelists and other experts on the subject have provided.

From my travels around the country, speaking with and listening to thousands of women—I've also gathered many anecdotes. In sharing these, I've tried to leave the conversational flavor of the questions and comments intact, just as I heard them.

More important, I want this book to take the sting and stigma out of menopause so we can talk about it openly, without feeling that if we say the word aloud, we're talking about *the end* of anything. We're not. We're talking about a new beginning, and we wouldn't trade our knowledge and life experience for anything. No, we don't want to be imposters of youth, because we can see the beauty of every age.

I learned long ago that at twenty we don't care what the world thinks of us. At thirty, we care very much what the world thinks of us. And at forty, we realize that the world wasn't thinking of us at all. Well, I want to add that at fifty and beyond, we need to change our focus. We need to discover what *we* think of the world and then to pursue the goals and activities that we choose to the fullest extent that we can.

This is our time! Let's enjoy it!

150
Most-Asked
Questions About
Menopause

CHAPTER 1

A Gathering of Women

At the turn of the last century—not so very long ago—women's lives generally ended at just about the same time as their reproductive lives. Those who lived beyond that period were considered to be "old." They even considered themselves to be old and were accepting of the miseries of old age.

Not so anymore. Today, statistics show that a woman who is healthy in her fifties will probably live into her mid-eighties. Every woman wants to have a good and active life during those bonus years. So women, in increasing numbers, are gathering together wherever good information is available about the midlife changes caused by menopause, that universal rite of passage.

And women are asking questions, hundreds of questions. They want to know how to stay in control during the transition from their reproductive years to their nonreproductive years, which is simply what menopause represents. This book contains the one hundred fifty most-asked questions at the educational programs at which I spoke and the good answers that were provided by the experts. Those questions and answers start in Chapter 2 and continue throughout the remainder of this book. Before we get into them, however, let's look at the programs themselves and at some new information on the subject of menopause.

These women's health programs deal with how to stay in control at midlife, the facts about menopause, and the facts about estrogen replacement therapy (ERT). The programs have been held in many cities throughout the United States. They've taken me from coast to coast—Arizona, California, Colorado, Louisiana, Pennsylvania, Minnesota—and I'm still "on the road." Women are attending these meetings in ever-increasing numbers. In Minneapolis in the summer of 1991, more than a thousand persons attended; half that number were expected. The ballroom of the hotel was filled to capacity and I was told that more than two hundred cars were turned away because the parking capacity had been exhausted, as well. I was delighted to learn that the North Women's Center, which had sponsored the program, gave a repeat performance some weeks later for those who could not be accommodated.

There have been other significant changes in the attendance at these lectures. Previously, only women were present. Lately, men, as a logical extension of their presence in childbirth classes, in the delivery room, and as "sharetakers" in the care of the children and of the home, make up anywhere from 10 percent to 20 percent of our audience, and their presence continues to grow. That's appropriate, because menopausal changes are family matters.

Women appear to be seeking information about menopause at an earlier age. The average age of the women in our audiences is now forty-seven years old and it seems to be dropping: Women in their thirties are beginning to show up. That's appropriate, too, because that's the age when you can start to make vital changes in your life-style concerning your diet, your exercise program, eliminating abusive substances such as cigarettes, and begin gaining skills for reducing negative stresses in your life wherever possible.

In all, more than ten thousand women and men have gathered at these programs to hear about menopause and to ask questions. They are asked to write their questions on index cards. These are collected prior to the lengthy question-and-answer period that concludes each program. Guests are also given an evaluation form to let us know their vital statistics (such as age and reproductive stage) and to tell us whether the programs have been helpful to them.

Many very interesting facts emerge from these index cards. Surprisingly, there are almost an equal number of questions concerning the basics of menopause as there are questions about more esoteric or special concerns. So questions such as "What is menopause?" and "When does it occur?" appear as often as "Does estrogen replacement therapy cause cancer?" and "I've had breast cancer; can I safely take estrogen to help me get through these hot flashes?"

In this book, I have selected the one hundred fifty most frequently asked questions and organized them so that you can find information easily, whatever your degree of familiarity with the subject. But, I caution you: This book offers information in an easy question-and-answer format, but it does not replace the good information that you should obtain from your own physician. It is meant to make you an educated consumer of health care related to menopausal symptoms, but only your own doctor has the final word on these subjects, because only he or she is familiar with your particular medical history.

For example, I remember standing in an enormous assembly hall in one large midwestern city listening carefully to the comments of the women who had drifted down to the podium to ask questions after the program concluded. One question stands out in my mind. It was asked by a beautiful young blond woman, who said, "I went through a natural menopause last year when I was thirty-seven. Does this mean that I'm fifty?" Try to answer that one! I asked her if she was on HRT. When she said, "Yes, I take estrogen and progestin," I told her "No—you are just thirty-seven and on hormone replacement therapy." Her equally attractive sister was standing next to her and said, "I'm thirty-five now. Will the same thing happen to me?" Naturally, I asked whether they knew their mother's history with menopause, because that could be an indicator of what their own experience might be. They told me that their mother had had a surgical menopause in her very early thirties and that, therefore, they did not know how her natural menopausal process might have occurred. Obviously, in this instance I couldn't offer any help, but I did suggest that these young women take these important questions to their gynecologist, which they had not done. Even though she was on hormone replacement therapy, the thirty-seven-year-old had not asked her doctor the question she had asked me. I hope both of these young women got their questions answered to their satisfaction.

Several months later, in Minneapolis, a woman came up to me after the program to ask whether I really knew how hard it is to get a doctor to take these midlife complaints seriously. I assured her that I did and that it worried me. That's why I continue to stress the forging of a patient/physician partnership—a liaison from which both partners will benefit. A knowledgeable patient is a joy to most physicians, because the doctor can render good care in a relatively short, easy visit and can feel fairly assured that the informed patient will comply with the decisions and directives that result from the visit.

Finding the right doctor takes serious and strenuous pursuit on the part of each woman. You should be looking for a doctor who does not consider the care of women at midlife to be his or her way to slow down toward retirement, but rather as a challenging and rewarding area of medicine in which to practice. Each woman needs to permit herself the liberty of asking her doctor this question directly: "Are you interested in the care of the menopausal woman?" And be prepared to discuss the situation. Perhaps your doctor is thrilled to deliver babies and is fascinated by work in the field of infertility, but is only lukewarm about the changes that can occur with menopause and the challenges of prescribing the right kind of HRT or ERT. In such an instance, you may need to ask your doctor to recommend another physician whose interest is more in line with your needs. No harm done and probably a happier situation for physician and patient alike.

That bit of advice reminds me of the gentleman who came up to me after a program in New Orleans and sort of hung around until the crowd thinned. He finally seemed to summon the courage to ask what I thought he could do to make his wife admit that something "strange was going on with her." He told me that her mood swings caused arguments that were alienating their children and that her headaches often kept her home from work and away from social activities. He had witnessed her hot flashes and felt the effect of her night sweats, but she continued to look for some remote medical or psychological problem rather than take her symptoms to the doctor. He didn't discuss their sex life with me, but I would be willing to bet that it had diminished or disappeared.

It took real love and concern for this man to show up

at our program in a sincere effort to get the facts so that he could help his wife, his children, himself, and his marriage. Why, why do we women deny? I think some of us don't want to know the facts. Others simply choose to ignore them, as I did. It seems that for some of us, any problem is better than the natural process of aging.

When I was invited to participate in a book fair in Ohio, two interesting things occurred that demonstrated the family involvement in menopause. Along with the eighty other participating authors, I was seated in a booth, which was set up to offer my book for sale. Several times that day young women came to buy the book for their mothers, and their comments were fascinating and revealing. A couple of young ladies had been asked directly by their mothers to pick up the book for them. But many of the other young women had read about the menopause book in the fair's publicity releases and were buying it on speculation. They were hoping to get their mothers to read it. I did a lot of informal counseling that day. I also found the behavior of one attractive middle-aged couple to be very interesting. They circled my booth several times, in deep discussion, occasionally glancing up at me. Finally, as the fair began to wind down and the crowd began to dwindle, they stopped to talk to me. The woman explained hesitantly that her husband was really the one who wanted her to buy our book. I smiled and hurried with my inscription as I sensed that she wanted to get away from the menopause booth quickly. Then, as I handed her the book and wished her good luck, she turned back to me and said passionately, *"I need it."* Her final request of me was for a bag to cover the beautiful purple volume that had the word *Menopause* emblazoned on its front and spine. That encounter made me wonder how long menopause would have to be treated as an undercover subject by some people.

As I tried to gain perspective on why this particular rite of passage was so threatening to women, I remembered that we women have fought long and hard for equality and that aging, as well as some of the more damaging symptoms of menopause—forgetfulness, nervousness, hot flashes—may feel as unbecoming to us in the workplace as they do in our family and social situations. Why is it that to this day, an aging man, graying at the temples, is labeled "dignified," while a woman with salt-and-pepper hair is considered "old"?

Americans have never rewarded aging, although the elderly are venerated in other cultures, such as in parts of India, where aging and wisdom are often synonymous. Now, however, demographics show that America is graying. That large group of postwar babies who were born between 1946 and 1964, nicknamed "baby boomers" by marketers, are quickly moving into the menopausal age range. Current advertising is being increasingly directed toward the baby boomers at midlife. An article in the business section of *The New York Times* (August 27, 1991) revealed that marketers have shifted their focus from "yuppies" to "grumpies"—that's "what demographers are calling the grown-up mature professionals. So, for the first time, the 78 million Americans, born between 1946 and 1964, are being courted because of their encroaching middle age." Just as the boomers have changed many aspects of our cultural, social, and business fabric by their sheer numbers, so, too, will they alter significantly the respect awarded the experience of age. It's about time!

Women often ask me if men go through menopause. The answer is no. Men do lose hormones just as women do, but they lose them slowly and much later in life. There is nothing in the male physiology that compares with the abrupt loss of estrogen at menopause, nor with the many body processes affected by that loss.

A 1991 Gallup study consisting of 1,410 interviews with both men and women revealed some new and interesting information about attitudes toward menopause. The large majority of the 705 women interviewed (all married and between the ages of forty and sixty) believed that they had a thorough knowledge of menopause and knew what to expect. Most of the Gallup respondents did not express anxiety about the onset of menopause. In fact, almost half of the women interviewed were looking forward to it! Just think: no more pregnancy worries! Many of the premenopausal women (40 percent) agreed. It is also interesting to note that most of these women did not express concern about the effects of menopause on their sexual relationships with their mates.

Men's expectations, however, were quite another matter. The husbands polled—and they were not the husbands of the women polled (only one person in any household was polled)—were less likely to say that they had a thorough understanding of menopause and were more likely than women to express concerns about the effects of menopause on their sexual relationships with their wives.

Most women in the Gallup telephone survey (96 percent) agreed that menopause is a normal state in a woman's life cycle. However, approximately 75 percent didn't accept the idea that menopause would change their lifestyles or that they must suffer without taking action. The husbands questioned were somewhat less positive. Forty percent believed that menopause would change lifestyles, no matter what women did about it. They were also less sure that their wives could avoid the physical symptoms of menopause.

The questions concerning sex showed interesting and positive results. Only 10 percent of the women surveyed thought that "after menopause a woman's ability to enjoy sex is greatly reduced." Once again, the men weren't so sure—though they appeared to question that assumption. This indicates to me that we women must educate the men in our lives, as well as ourselves.

From the studies and questionnaires I reviewed and the women I talked with, I found that the college-educated respondents believed that they had the facts and felt fairly comfortable about issues of control—believing that they could help themselves by monitoring their diets, exercising more, and taking better care of themselves in general. Yet when asked specific questions about estrogen replacement therapy and its role in protecting and prolonging life, these women did not have a clear picture of what ERT can and cannot do or who can and cannot take it. So although the women polled seemed long on their optimism and determination to make menopause a nondisruptive force in their lives, they fell short on having the state-of-the-art facts and knowledge about what to do to help themselves. It is that kind of information that led me to write this book.

Finally, it seemed clear from the women surveyed that those who had a comfortable relationship with their physicians—who felt that their doctors were individuals with whom they could communicate—had a better understanding of menopause in general as well as of the long-term benefits of ERT.

These Gallup survey results are important because they show an increased awareness of menopause on the part of both the men and women studied and a higher incidence of knowledge than a Harris telephone study done in 1987, in which five hundred women surveyed in ten major cities across the United States clearly showed confusion about menopause and its treatment. This indicates that new and increased information about menopause, coupled with a more general acceptance of talking about it, is helping women and men alike.

The Gallup survey appears to contradict the information provided by the women and men who attend the educational programs, but let's look at the differences carefully. The women and men who attend the programs are self-selected: They learn about the free program and *choose* to attend; some choose to ask questions and fill out the evaluation forms. In contrast, the women and men surveyed in the Gallup study, through the use of standard random-digit dialing technique, are selected at random. Yet, they do not always trust that their identities are not known (they *are* unknown) and they may be less comfortable giving information over the telephone to faceless strangers, particularly about their attitudes toward sexual matters. I believe this disparity is understandable, but the comments from both sources are interesting nevertheless.

You can start to see the mainstreaming of menopause all around you. For example, on television Barbara Walters (a woman of a certain age herself) devoted a *20/20* segment to menopause. Reporters, newscasters, and talk-show hosts have become more comfortable in reporting menopause news, covering the results of each new study concerning estrogen replacement therapy, particularly those concerned with its effects on heart disease (see Chapter 9 for more on this subject). And some of our most frequently watched television situation comedies have dealt with menopause. It wasn't easy for me to hear the end-of-productive-life concerns of her family when Clair Huxtable, on *The Cosby Show*, made her midlife transition, and even sadder when Mariette Hartley, on the short-lived sitcom, *WIOU*, mistook the symptoms caused by menopause for the symptoms of late pregnancy. That episode ended with her in "end-of-productive-life" tears. So we're talking and we're clearing the air and we're loading the airwaves with menopause talk.

Menopause has even crept into some of the best-selling books of the past couple of years, although mostly in a negative manner. For example, at the denouement of P. D. James's *Devices and Desires*, when the killer is admitting her crime, she explains to her closest friend how the police will view her: "They'll just see me as a postmenopausal neurotic woman gone temporarily off her head," she says.

In Scott Turow's recent book, *Burden of Proof*, the protagonist, lawyer Allesandro Stern, has sex for the first time after his wife's suicide and he and his partner are talking in the afterglow. What are they discussing? Right! Menopause.

In Peter Furth's book entitled *An American Cassandra*, about famed journalist Dorothy Thompson, he writes of the well-known quote of Alice Roosevelt Longworth, who said, to explain Thompson's tartness, "She is the first woman in history who has had her menopause in public and made it pay." Also from the very public political arena comes a story in *Keeper of the Gate* by Selwa "Lucky" Roosevelt, former President Ronald Reagan's chief of protocol. In it she talks of her dinner in Bangkok with the king and queen of Thailand in which the queen, explaining her absence from the world political/social scene, says to Lucky, "I have fought the Communists all my life, but I cannot fight the menopause."

Even though these references are not menopause-positive, they do indicate that menopause is being recognized and discussed openly at last!

These are just a few of the many places and ways that menopause is becoming a socially acceptable and useful word. The importance of that acceptance is vital to all of us. It empowers us with the strength and knowledge to become our own ombudswomen—to find the doctor most helpful to us—and it gives us the savvy to engage in life-styles that may reward us with not just more years, but more good years.

"To live effectively is to live with adequate information," wrote Norbert Wiener in 1954 in his book *The Human Use of Human Beings*. With the recent explosion of information about menopause and its effective management, those words are especially true for women today.

CHAPTER 2

What Is Menopause?

While touring the country doing lectures about my experience with menopause, I began to collect some of the personal stories that women shared with me, particularly those told during that warm time of good conversation when women gathered around after the formal part of the program had ended.

Often we discussed our mother's experiences with menopause. My mother—who was a terrific communicator—taught me nothing about menopause. Maybe she suffered in silence or whispered about it to her friends, but I didn't learn anything. I frequently ask women if they remember their mothers' menopausal experiences. The answers range from "I remember that she seemed to act weird for a long while—maybe that was during her menopause" to: "She seemed despondent, depressed, more nervous than I had ever seen her"; "She cried a lot"; "She seemed to become reclusive"; and "She grew old suddenly before my eyes." Many women I spoke with didn't even have this much of a clue, as their mothers and grandmothers had died before this signal event had occurred.

Therefore, many women are left with lots of questions. As one lovely lady confided to me in Pittsburgh, "I never could have brought myself to talk about *that* with my doctor—it's too personal!"

At the programs I have heard as many questions about the basic process of menopause itself (what is it, when does it occur, how will I know the symptoms, etc.) as about the more complex issues of hormone replacement therapy, osteoporosis, heart disease, and cancer. There are remarkably few questions about sex, and that always surprises me. As with the Gallup survey, that may be because people continue to feel reticent when it comes to asking questions about sex.

In this chapter, I want to share those rudimentary questions and answers so that we are basing our understanding of menopause upon the same definitions and facts. The questions below have been asked at every program. These questions and answers will serve as an excellent foundation for all of the other questions and answers in this book.

1. CAN YOU TELL ME EXACTLY WHAT MENOPAUSE IS?
Menopause is essentially the end of menstruation, that cyclical ovarian process which began in puberty. Doctors usually consider your menopause to be complete when you have not had a period for one full year. This rite of

passage carries you to your nonreproductive stage of life, marking the end of your childbearing years.

2. Why is it called menopause?

The word *menopause* comes from a monograph written in 1812 by C. P. L. Gardanne, a French gynecologist, which dealt with the change of life. In it Gardanne combined the Greek words meaning "month" and "end" and coined the word *ménépausie*.

3. When does menopause occur?

The average age for menopause is fifty-one years, four months. This fact has barely changed over the years. On average, menopause occurs just beyond the midpoint of the ten-year period from age forty-five to age fifty-five. This period of years, called the *climacteric*, generally brackets the the last decade of the menses. As with any average, however, there are exceptions. There are some women who experience a natural menopause as early as in their thirties, like the young midwestern woman whom I described in Chapter 1. On the other hand, women also have been known not to experience menopause until their sixties, though that is not common.

4. Isn't menopause a process that can take years to go through?

Yes, it can be. In your thirties, your periods may begin to change, arriving closer together, further apart, or skipping a month altogether. The bleeding may get heavier, or rarely, lighter. This can be an unpredictable and confusing time. In *Managing Your Menopause*, my coauthor, Dr. Wulf H. Utian, revealed his theory that sudden changes in the menstrual cycle or sudden incidence of PMS, when it begins in your thirties, may really be pre*menopausal* syndrome, rather than pre*menstrual* syndrome, caused by the aging of the ovary and its slow depletion of eggs.

Menopause is technically the term that describes your final period. There are other terms used to describe these years of change. *Perimenopause* (hereafter referred to as premenopause) literally means the years before menopause and is described by Dr. Lila Nachtigall in her book *Estrogen: The Facts Can Change Your Life* (coauthored with Joan Heilman) as the time when you gradually stop ovulating and your ovaries slow their production of the two female hormones estrogen and progesterone. *Postmenopause* refers, of course, to the years after a woman's final period, but before she becomes elderly.

5. BESIDES A CHANGE IN MENSES, WHAT ARE THE OTHER EARLY SYMPTOMS OF MENOPAUSE?
The most common (and one of the first) symptoms of menopause is the hot flash, which occurs to a greater or lesser degree in 75 percent of menopausal women. Other symptoms include night sweats (hot flashes that occur at night, sometimes waking you), palpitations, insomnia, disorientation, mood swings, and minor depression.

There is a long list of other symptoms, common and uncommon, and a great difference in symptomology among women. That's because although menopause is universal, it is also unique to each woman. So if you sit down with a group of three or four friends to discuss menopausal symptoms, it is likely that you each may have different symptoms or you may have the same symptoms to different degrees. Symptoms will be discussed in greater detail in Chapter 3.

6. I'M FORTY-EIGHT AND HAVEN'T HAD A PERIOD IN THREE MONTHS. I HAVE NO OTHER SYMPTOMS. AM I IN MENOPAUSE? SHOULD I SEE A DOCTOR?

Yes, see your doctor: That is the only way to learn what is happening in your body. If your physician suspects menopause, he or she will probably do a simple blood test to determine either the amount of follicle stimulating hormone (FSH) or the serum estradiol concentration. Those numbers should indicate whether you are in menopause. Produced by the pituitary gland located in the base of the skull, FSH causes the release of the egg from the ovaries. If it is working but can't accomplish that release—as would be indicated by a higher FSH number—it is a signal that the ovary is running (or has run) out of eggs and is ceasing its production of the female sex hormones, which is what causes menopause. The newer serum estradiol concentration test tells even more about how much estrogen is in our blood.

7. CAN A WOMAN BE GOING THROUGH MENOPAUSE AT AGE THIRTY-NINE?

Thirty-nine is on the early side, but of course, she can. This is an excellent time to see your physician to determine whether you are having menopausal symptoms. Approximately eight out of one hundred women may go through premature menopause.

8. WHAT IS PREMATURE MENOPAUSE?

Menopause is usually considered to be premature when it occurs in a woman's thirties or early forties. Menopause is also called premature when it occurs even earlier following surgical removal of the ovaries because of disease, such as ovarian cancer or endometriosis.

9. SHOULD PREMATURE MENOPAUSE ALWAYS BE TREATED?

Most physicians believe that it should be treated with hormone replacement therapy. Without treatment, a young woman going through menopause will be deprived of the protective effects of estrogen in her body many years earlier than she would have been naturally. This puts her at greater risk for thinning, porous bone and osteoporosis, for example, which is discussed in Chapter 8.

10. IF MY MOTHER HAD A DIFFICULT MENOPAUSE, WILL I HAVE A DIFFICULT TIME, TOO?

This question is difficult to answer definitively because

"bad" or "difficult" are subjective evaluations. Heredity does play a large role in the menopause. A good rule of thumb is that if your mother had an early menopause, it is likely that you will, too. However, there are other factors to take into account. For example, if your mother smoked cigarettes, her menopause may have been accelerated by as much as five to ten years. The reverse is true, too: If you smoke and your mother did not, your menopause may occur earlier than hers did. The age at which you had your first period, however, is not believed to determine the age at which you will experience menopause. In addition, your life-style compared to your mother's, your body build, and your unique life stresses may be contributing factors to the ease or difficulty of menopause.

For example, at the average age of menopause, in our early fifties, we are dealing with many other significant changes in our lives. Our children are leaving home or have left, we may become grandparents for the first time, our own parents may be ill and in need of our care, and our work and home duties may change in other important ways. With divorce claiming one out of two marriages, a more recent development is that many more of our children are returning to the nest as single parents, bringing their children with them. For the growing number of women who have delayed childbearing until their late thirties and forties, dealing with teenage offspring when menopause occurs can be replete with its own set of problems.

The way in which our culture regards aging may affect symptoms as well. For example, women in the Rajput class in India, where aging is rewarded as a time when women may join the men in positions of power and rule-making, menopause may be welcomed and the symptoms disregarded. In a youth-oriented culture such as ours, aging may be dreaded and symptoms may be experienced as being bothersome and embarrassing.

11. IS THERE ANYTHING YOU CAN DO IN YOUR TWENTIES AND THIRTIES TO HELP LESSEN NEGATIVE MENOPAUSAL SYMPTOMS?

First of all, educate yourself about the changes your body will go through and what can work to help you make those transitions easily. There are no guarantees, but studies show that building peak bone mass in your younger years by eating calcium-rich foods and exercising can help protect your bones from osteoporosis later in life. (Chapter 8 covers questions about osteoporosis.) So it's never too soon to begin enjoying a low-fat, high-fiber diet and to incorporate the three kinds of exercise you need—flexibility exercise, weight-bearing exercise, and aerobic exercise—into your life. The exercise you choose should be varied enough to hold your interest and you should engage in it for a minimum of twenty to thirty minutes three times a week, once you are in shape. These matters and their effect upon you, your heart, and your bones are discussed in Chapter 13. Furthermore, if you smoke, quit! Smoking is harmful to all of your body's systems. Some theories suggest that nicotine creates metabolic changes that may affect the ovary; others that cigarette smoke causes the liver to destroy estrogen. In either instance smoking's effect on the ovaries can lead to an earlier menopause. It also may cause thinning of bone by decreasing estrogen production, which may block calcium absorption. Smoking also is a well-researched culprit in heart disease and certain forms of cancer. Finally, keep your caffeine consumption to a minimum—no more than two cups (or less) per day—and alcohol intake to just one cocktail or one glass of wine or beer daily.

Another good idea would be to discuss menopause with your mother and your grandmother, if they are still living. Find out how menopause affected them, how early its onset was, and what they did, if anything, to offset their symptoms.

12. Is MENOPAUSE DELAYED BY CONTINUOUS CHILD BEARING?

Although menopause symptoms may be lessened by having had pregnancies, the number of pregnancies usually does not delay menopause. A late menopause may, in rare instances, permit late pregnancy.

13. DOES A LATE MENOPAUSE RELATE TO A LONGER LIFE?

A late menopause means that your ovaries are still producing a strong supply of estrogen. This may enable you to look younger on the outside and not be subjected to the thinning and drying of the tissues on the inside that are characteristic of estrogen reduction. Your bladder and vagina may retain their moisture and muscle tone and your bones and heart may be protected for a longer period of time. However, according to obstetrician-gynecologist and coauthor of *Menopause* (with Dr. Lindsay Curtis) Dr. Mary K. Beard, with whom I sat on a panel in Minneapolis, there are disadvantages to late menopause as well. Besides causing you to have to deal with continued periods and with preventing an undesired late pregnancy, there is also a slight risk of ovarian cancer. This is why even a late menopause should send you to your doctor for an annual physical and pelvic examination so you can enjoy the benefits and eliminate the risks of a late menopause.

14. I'VE BEEN ON BIRTH CONTROL PILLS FOR EIGHTEEN YEARS, SO HOW DO I KNOW IF I AM GOING THROUGH MENOPAUSE AND WHEN?
You did not indicate your age, but according to most authorities, women should consider switching to other means of birth control after the age of forty-five. Neither the FSH test nor the serum estradiol concentration test is a reliable indicator of reduced estrogen production if you are on oral contraceptives. Discuss this matter with your doctor.

15. WHAT IS "POSTMENOPAUSE"? HOW DO YOU KNOW WHEN MENOPAUSE HAS BECOME *POST*MENOPAUSE?

Postmenopause is defined as occurring when ovarian function has ceased and continuing until old age. Menopause is thought to be over when its symptoms have abated. The years that follow are considered to be the postmenopausal years. Because each woman's experience of menopause is unique, I know of no clear lines of demarcation to mark the end of menopause and the beginning of the postmenopause.

Those years that follow menopause can be the very best years of your life. As I write these words, I am reminded of several lines I read years ago in the Afterword of *Passages* by Gail Sheehy:

> *The greatest surprise of all was to find that in every group studied, whether men or women, the most satisfying stages in their lives were the later ones. Simply, older is better.*

CHAPTER 3

What Are the Signs and Symptoms of Menopause?

The average age at which menopause occurs has remained constant for centuries, hovering around fifty-one years, four months. What *has* changed is women's life expectancy. Today, statistics show that women who are healthy in their fifties generally can look forward to living well into their eighties—they have a full one third of life left to live! That's especially good news for those of us who think ahead and plan well.

Thirty-five is a good age to focus upon. That may be the optimum time to get back into the health care delivery system, if you've drifted away, and to get serious, if you're not already, about your exercise program, your nutrition (including calcium consumption), and about dropping bad habits such as smoking (I'll be answering questions about those health habits in Chapter 13).

Thirty-five is also a perfect age to begin educating yourself about the signs and the symptoms of menopause and other age-related changes, although most of the women attending our programs are between forty-five and fifty-five years old. Many women in their sixties and seventies are in the audiences as well. These women tell me that our talks provide useful information that they can still incorporate successfully into their life-styles. In other words, it's never *too early* to learn about menopause and it's never *too late*, either. Well, almost never. I had a really heartbreaking experience shortly after our first book was published. A sharp young reporter from a first-rate daily newspaper in a small community telephoned me for an interview. I'll call her Carol. Carol was so deeply interested in my personal story and asked such penetrating questions that I couldn't help asking her age. I wondered why someone who was just twenty-six years old would be so intensely involved in something that happened to me when I was twice her age.

We talked for a long time. Slowly and subtly, our roles reversed and I became the interviewer. It was then that I learned to my great sorrow that her mother had committed suicide recently, at the age of fifty-two. Carol told me that prior to the year just past, her mother had always been in good mental and physical health and in good spirits. Having raised a large family and taken care of her husband and home, she also had a job outside the home and had been a bright and eager community activist. Then suddenly she wasn't the same anymore.

Carol, graduating that spring from a large midwestern university, came home the Easter before graduation and knew quickly that her mother was different. She frequently saw her mother perspiring and then, at other times, shaking with chills. Her mother appeared to be very nervous and often was quarrelsome. She walked the halls at night or sat on the edge of the sofa in the living room staring into space long after the family was asleep.

Carol's mother was on a collision course. The family doctor of many years felt he couldn't help her and sent her to a psychiatrist. The psychiatrist prescribed anti-anxiety pills and sedatives, which didn't seem to help. Thus, when the psychiatrist recommended institutionalizing Carol's mother so that her condition could be closely monitored and therapy could be provided, his suggestion seemed to the family like a reasonable course of action. After the first six weeks of living inside the institution, when she seemed to be quieter and more accepting of her new strange and remote self, the doctors felt that Carol's mother could start to go home on weekend visits. The first weekend she was home, after writing a long letter in which she expressed hopelessness about her condition and her future and made clear her commitment not to be a burden to her family, she committed suicide.

Carol had read in our first book my account of the eight hellish weeks I endured before it was realized that I needed hormone replacement therapy—not tranquilizers, sedatives, or psychotherapy—to bring back my old self. She connected this story immediately to her mother's sad demise. Like a good reporter, she interviewed me in a series of phone calls. She had already begun to try to interview everyone involved in her mother's care. Most doors were shut to her as she tried to obtain her mother's medical records. She could find no record that her mother had ever been evaluated for the hormone deficiency associated with menopause.

Carol will never know for certain what events or conditions conspired to drive her mother to end her own life. I cry for Carol and for her need to search for clues to this tragic mystery. Who can know what really happened?

I only know that if Carol's mother had been aware of the symptoms of menopause, or if Carol or her father or any other member of that closely knit family had had a knowledge of the symptoms caused by estrogen loss, perhaps the outcome would have been different. It's a big perhaps, but Carol clings to it.

After her stirring article about our book appeared, Carol and I talked again and agreed that we both felt better for having alerted the public to what menopause can be like when it's bad and that it can be bad for 15 percent of women. Perhaps, through reading her article and our book, women experiencing odd and inexplicable changes at midlife might relate their symptoms to the symptoms of menopause and would check them out with their doctors. Further, I hoped that family members, aware of the signs that some women demonstrate at the time of menopause, might be able to help.

I've already explained that although menopause is universal, its symptoms are also unique to each woman. Women's questions about symptoms are as varied as the symptoms themselves. Many of the questions asked at the programs are personal and patient-specific. I suggest that women direct these questions to their physicians. Only following the physician's taking of a complete history and performing a physical examination and whatever tests the doctor deems appropriate can individual questions be answered. The following questions are the most-asked non-individual-specific questions that I encountered about the symptoms of menopause.

16. WHAT ARE THE MOST COMMON SYMPTOMS OF MENOPAUSE?

A change in your period is the most common premenopausal symptom. Periods can get closer together, further apart, or heavier in flow. Next, according to the 1991 Gallup survey, the large majority (87 percent) of menopausal and postmenopausal women studied reported that they have experienced hot flashes (this percentage is higher than the 75 percent usually noted), and more than two thirds (68 percent) say that they have night sweats. At least three in five women (61 percent) report that they have experienced anxiety, irritability, or nervousness, and nearly as many (58 percent) have experienced mood swings or some depression. With the exception of hot flashes, women currently going through menopause are more likely than postmenopausal women to say that they have experienced each of these conditions. This may be because they were going through the phase most abundant in symptoms at the time of the interview.

Other symptoms that were sometimes mentioned include changes in urination—retention, urgency, frequency—food cravings, fluid retention, and insomnia. A whole grab bag of other problems are reported, but many of them seem to be secondary effects of other symptoms. For example, painful intercourse could be the result of drying and thinning vaginal tissue. Lack of interest in sex can also be a result of a lack of estrogen. At a recent program, one woman explained, "I go to bed in scanty nightgowns with the fan going; my husband wears a sweatshirt."

17. WHAT EXACTLY IS A HOT FLASH?

The hot flash is a vasomotor symptom caused by a change in brain chemistry. This chemical change is the result of an abrupt drop in the body's production of estrogen. It works somewhat like the domino theory. The chemical changes in the brain affect the temperature control center in the hypothalamus gland, a part of the brain. The hypothalamus causes the release of hormones that direct a decrease in the body's core temperature set point. That decrease, in turn, causes the dilation of the blood vessels of the skin and sweating as your body—accustomed to its own set point—begins working to reset its own thermostat. This entire disturbance makes you feel as if your internal thermostat has gone haywire.

My hot flashes usually began in the traditional way, starting just above my waist, and then spreading quickly upward to envelop my chest, back, neck, face, and scalp. They were infrequent, but brutal. Some women have as few as two or three per day, while others may have as many as fifty flashes per day. Some women are barely disturbed by their flashes; others feel debilitated by them. There are reports from women who had hot flashes for only a few months and from others who endured flashes for years. Stress, alcohol, caffeine, and even spicy food seem to set off a hot flash in some women. Some women report cold flashes, as well. Flashes can create "the thermostat wars" in some households.

18. IS THERE A RELATIONSHIP BETWEEN PREMENSTRUAL SYNDROME (PMS) AND MENOPAUSE?

Widely differing opinions exist on this point. Some physicians say that early PMS that begins in the teens, with its symptoms of headaches, heavy menstrual flow, and severe cramping, may diminish around the time of menopause; others suggest that PMS may actually worsen as the premenopause approaches, and then coalesce with it. There appears to be no medical consensus that the more PMS problems you have, the more trouble you'll have with menopause. Today, PMS is believed by some doctors to be an abnormal brain response to normal hormone levels.

There is some medical agreement that women who develop PMS symptoms or who experience a worsening of existing PMS symptoms for the first time after the age of thirty may be demonstrating the first change in their hormone balance as their bodies prepare for menopause. As with so many women's health issues, more research is needed.

19. IS THERE ANY WAY I CAN FIGURE OUT WHETHER I HAVE PMS?

If you keep a calendar of your PMS symptoms, you'll learn a lot about your body's unique response to monthly hormonal changes. Note the date when each of your premenstrual symptoms begins and how long it lasts. Include any changes in your symptoms as they occur. Remember, you have a lot of company in trying to understand and work through PMS: Some seven million women suffer from PMS. For some women, PMS lasts just two or three days a month; for others it can last up to the whole fourteen days before your period. Studies show that PMS usually enters your life with your first period or after your first pregnancy. My advice is: Don't suffer in silence as I did. I came from the generation in which this monthly disturbance didn't even have the dignity of having initials! Today, doctors are there to help you deal with PMS. See your doctor and take your partner with you. Aerobic exercise, dietary changes, limiting caffeine, chocolate, and cheese may help, too.

20. IS BREAST TENDERNESS COMMON DURING MENO-
PAUSE, AND IS IT A SYMPTOM OF MENOPAUSE?
Breast tissue is very responsive to estrogen and progesterone that is produced during the normal menstrual
cycle. Tenderness in the breast is quite common in menopause if you are on hormone replacement therapy. Just
as breast changes such as tenderness and enlargement
may have occurred just before your period, so it is not
uncommon to experience discomfort from fluid retention
and stimulation of the mammary glands if you are on
hormone therapy. This symptom should diminish or become tolerable in a few weeks after beginning therapy.
If it does not, see your doctor. Dosage adjustment may
be the answer. Breast examinations and mammography
questions are covered in more detail in Chapter 7.

21. IS AN INCREASE IN FACIAL HAIR TO BE EXPECTED IN
MENOPAUSE?
It can happen. This happens because once we run out
of estrogen, the androgens, or male hormones, that we
all produce naturally are no longer "opposed," that is,
kept in balance, by our estrogen. Hair starts to grow in
response to the same level of male hormone that was
always there, but (because of estrogen's opposing influ-
ence) didn't cause hair to grow. This may make hair seem
to grow in a more masculine manner, such as on the
upper lip, chin, and cheeks. If you should decide to take
estrogen replacement therapy, the male-pattern hair
growth stops, but unwanted hair that is already there

remains. Hair removal can be achieved in many ways, but electrolysis is the only permanent method of hair removal. Other temporary methods include shaving, tweezing, waxing, or the use of depilatories.

22. IS IT NORMAL TO BE LOSING HAIR ON MY HEAD, ARMS, LEGS, AND PUBIC AREA?

Yes, hair loss can be a result of body processes that are changed by menopause. It works this way. The follicles for each of our hairs is located deep in our skin, or dermis. The hair follicle is supported by collagen-rich tissue. Your skin, like many other organs in your body, requires estrogen in order to function normally. As we age, the amount of collagen we make tends to diminish, making the skin thinner and drier; at the same time, the fat and muscle under the skin also shrinks. When these tissues which supported the hair follicles diminish, hair loss may result. Hair loss may be very stressful for you, but don't suffer in silence. Discuss hair loss with your gynecologist and with a dermatologist, who may be able to help.

23. WHAT URINARY SYMPTOMS ARE MOST ASSOCIATED WITH MENOPAUSE?

Once you have run out of estrogen, you may find yourself faced with frequent urinary tract infections, because the tissue of the urethra and bladder may become thinner and lose its elasticity and thus become more open to infection or irritation. This is a concern you should take to your physician immediately to make sure you do not experience repeated infections that could endanger your bladder or kidneys. Other urinary symptoms that are common causes of discomfort after menopause include a change in the frequency of urination, an urge to urinate that is so pronounced that you can barely hold it in until you get to the toilet, and what is to many women the worst of all: the inability to hold in urine under stress. Physicians refer to this condition as stress incontinence. A little urine may leak when you laugh, sneeze, cough, or run to return that tennis ball. A full description of how to do Kegel exercises, which may help, is in Chapter 10.

24. WHAT CAN CAUSE DEPRESSION IN MENOPAUSE, AND WHAT CAN BE DONE ABOUT IT?
Minor depression occurring around the time of meno-pause affects a significant number of women. It may be the result of night sweats (hot flashes that occur at night) which disturb healthy sleep patterns and which are fre-quently followed by chills and a need to cover and un-cover oneself (or even to change nightclothes and bedclothes in extreme cases). The fatigue caused by interrupted sleep may itself cause us to feel depressed. It is important, however, that these menopause symp-tom–related feelings of depression be differentiated from emotional illness. Hormone replacement therapy may enable us to sleep better and feel better and, thus, will

lift the minor depression caused by hot flashes, night sweats, insomnia, and other symptoms. However, it probably will have no effect on emotional illnesses. It is easy to confuse the depression, anxiety, mood swings, and irritability that may occur around the time of menopause with the same feelings caused by other environmental and psychological factors.

Earlier in this book, I related a long list of life changes that often occur at the time of menopause: the illness or death of one or more of your parents; children leaving home and leaving you with an empty nest; the stress of your job; other stresses such as a magnified lack of certainty about your life/work choices, or your mate's midlife crisis or work crisis; or divorce, which is becoming more common later in life. You can see how easily the environmental causes of depression can become confused with the symptom-related causes. So, again, ERT or HRT may be worth a try—and *if* your depression is caused by the symptoms related to a lack of estrogen, it should clear up fairly quickly once you begin replacement therapy. In either case, talk to your doctor! Some doctors say that estrogen has a "mental tonic" effect; others disagree.

25. CAN MY INSOMNIA BE RELATED TO PREMENOPAUSE EVEN IF I AM NOT EXPERIENCING HOT FLASHES?
Although 75 percent of all women experience hot flashes during menopause, other women do not. You may, however, suffer from insomnia caused by the same vasomotor

symptoms that cause hot flashes. In fact, if changes in your menstrual cycle are occurring along with sleep problems, these may be signs of approaching menopause.

26. ARE PALPITATIONS A COMMON SYMPTOM OF MENOPAUSE? WHAT CAN BE DONE ABOUT THEM?

That out-of-sync sensation you experience, as if your heart is beating too fast, may be the result of palpitations. These rapid and distinct heartbeats are vasomotor phenomena that are a fairly common symptom of menopause, but they also can be caused by many other factors such as too much caffeine or nicotine. More important, they can be a signal of heart disease, so discuss any palpitations with your physician.

27. IS FEELING FATIGUED A COMMON SYMPTOM OF MENOPAUSE?

On our questionnaires, fatigue is listed as one of the most common symptoms of menopause. Obviously, if you are suffering sleep deprivation as a result of hot flashes, you will feel tired. If you are experiencing stress, you may also feel fatigued. Before you decide that your fatigue is a menopausal symptom, it is important that you and your physician rule out any other physical or psychological reasons for your fatigue.

28. DOES STRESS INCREASE MENOPAUSAL SYMPTOMS?

Stress can make everything worse. If you're nervous, anxious, irritable, or somewhat depressed as a result of menopausal symptoms, other stresses just compound the problem. Lack of sufficient sleep due to stress or to hot flashes can make life's negative stresses worse. It's a vicious cycle: Decreased estrogen may make you more vulnerable to stress, and stress can exacerbate menopausal symptoms. Pinpointing the origins of symptoms is a job for the patient/physician partnership. Once you and your doctor have sorted out the sources of your negative stresses, be they social, environmental, or physiological, you need to work toward eliminating as many of them as you can. Chapter 13 will offer tips for stress reduction.

29. IS MEMORY AFFECTED BY MENOPAUSE?

Very few studies have been done on the effect of menopause upon short-term memory. Nevertheless, women themselves describe this problem in many ways, starting with difficulty concentrating or forgetting where they put things. For example, how often do you have to look for your car keys or your sunglasses, or struggle to recapture lost thoughts or the ends of your sentences? As one woman told me, "Having spent the first fifty years of my life getting to the point where I could function really well in my life, I really resent feeling mentally foggy now." Another added wistfully, "These days, I always feel like my thoughts are too little or too late." Two coping tips that are highly rated in our program questionnaires are making more reminder lists and notes for yourself and forgiving yourself for those momentary lapses. Remember, according to the experts, all of us may experience a slight decline in memory capacity as we age.

30. HOW LONG DO MENOPAUSAL SYMPTOMS LAST?

No one can give us a definite answer. Just as premenopause differs from woman to woman, so does the conclusion of menopausal symptoms. The symptoms disappear in a few years for some women; for others, symptoms continue much longer. Women who choose to take estrogen replacement therapy may experience no symptoms while they are on therapy. Today, many physicians believe that women can remain on estrogen for at least ten years, or even indefinitely. For many women on ERT, once they've gone beyond the number of years during which they would have had symptoms, no symptoms return when they discontinue ERT.

All in all, the symptoms of menopause are many and varied. Some, like the hot flash, rise up and announce themselves to us, demanding attention; others, like the slow thinning of our bones, move silently, creating no concern until they cause a crisis. A full chapter is devoted to osteoporosis, Chapter 8.

It is inevitable that the changes that occur within us at menopause cause other changes as well. That is why menopause represents the ideal time to take both care and control of yourself, and to bring to your health care a knowledge and understanding of your physiology at midlife. Our physiology and psychology is complex and miraculous. It is amazing that all of our systems are designed to act and interact so well.

In that best-selling classic in celebration of life, *The Lives of a Cell: Notes of a Biology Watcher*, author Lewis Thomas reminds us: Statistically the probability of any one of us being here is so small that you'd think the mere fact of existing would keep us all in a contented dazzlement of surprise.

CHAPTER 4

What Is Surgical Menopause?

Surgical menopause occurs as the result of a hysterectomy in which a bilateral oophorectomy is also performed. In plain terms, this means that your uterus and both of your ovaries were surgically removed.

As you know, menopause is caused by the ovary running out of eggs and ceasing its production of the female hormones estrogen and progesterone. When both ovaries are removed surgically, the primary source of the female hormones is suddenly gone. This incisive change will thrust you into menopause, unless your physician prescribes estrogen replacement therapy immediately to replace your lost hormones. One young woman told me that she left the operating room after this procedure with a transdermal estrogen patch already stuck on her abdomen, and she never experienced a single menopausal symptom.

The practice of replacing the lost hormones immediately is becoming more common for women who can take ERT and who choose to do so. (Assessing who may be a good candidate for ERT is discussed in Chapter 6.) This is particularly important for younger women undergoing hysterectomy and oophorectomy, since some studies show that along with the other menopausal symptoms that may occur, some young women may develop depression within three years after the surgery.

If a medical decision has been made that you must have a hysterectomy, feel free to ask a lot of questions. Find out if your ovaries will be involved in the procedure. It is important to try to save the ovaries, if they are healthy. Even one ovary, in most cases, will still work to produce estrogen and progesterone and hold off the sudden onset of menopause until normal menopausal age. The estrogen produced by the ovaries bathes and nourishes many different organs and processes in the body. Unless an emergency hysterectomy is indicated, try and learn as much as you can about the procedure and let your surgeon know that you are interested in preserving as much of your reproductive anatomy as possible. Read! Ask questions! Get a second opinion!

More than 590,000 hysterectomies are performed in U.S. civilian hospitals each year, and the median age of women undergoing this procedure is forty and a half. That is more than *five times* the number of hysterectomies that are performed in Europe. Hysterectomy is the most frequently performed operation on women in the United States. If an oophorectomy is performed at the time of the hysterectomy, premature aging begins unless steps are taken to prevent it. Dr. Winnifred Cutler reports in *Hysterectomy: Before and After* that the risks of hysterectomy with oophorectomy include well-documented increased incidence of heart attack, reduction in sexual functioning, and increased urinary incontinence. Speaking as someone who underwent a hysterectomy (leaving my ovaries intact) in my early forties, I personally recommend to you her book and Dr. Vicki Hufnagel's *No More Hysterectomies* before you undergo this all-too-common surgical procedure.

Questions about surgical menopause are quite common at the programs and usually revolve around the following issues.

31. HOW DOES SURGICAL MENOPAUSE DIFFER FROM NATURAL MENOPAUSE?

Natural menopause occurs when your ovaries deplete their lifetime supply of eggs and gradually shut down their hormone production system. The average age for this shutdown is about fifty-one years, four months, but the process begins many years before. In contrast, a surgical menopause occurs when the ovaries are surgically removed and its onset is sudden and full-blown.

32. I HAD A HYSTERECTOMY IN MY THIRTIES. I HAVE ONE OVARY LEFT. I AM NOW FIFTY. WILL I EXPERIENCE MENOPAUSE?

Yes, you will. The one ovary has obviously produced

enough estrogen for you to remain premenopausal, but you are now in the menopausal age range and can probably expect menopause shortly.

33. AFTER A HYSTERECTOMY (WITH BOTH OVARIES STILL INTACT), HOW CAN I TELL WHEN I AM GOING THROUGH MENOPAUSE?

Some women with hysterectomies sail through menopause, while others suffer through it just as some women with nonsurgical intervention would. Usually symptoms do herald the onset of menopause. You may experience hot flashes, night sweats, sleep problems, or vaginal dryness. If you have no overt symptoms yet, an FSH test or measuring your serum estradiol concentration (described in Chapter 2) will let you know whether or not you are menopausal.

34. AFTER A HYSTERECTOMY AND OOPHORECTOMY, WHAT CAN I EXPECT IN TERMS OF MENOPAUSAL SYMPTOMS?

A hysterectomy with bilateral oophorectomy usually brings on menopausal symptoms within a few days (or even hours) after the surgery, unless ERT is given. If you are one of the few lucky women who are not sensitive to or bothered by the loss of the sex hormones—estrogen, progesterone, and testosterone—you may skip some of the symptoms for a while or even forever. Your doctor may suggest one of the blood tests, as discussed in Chapter 2, to learn where you are in the menopause process.

35. I HAD A HYSTERECTOMY YEARS AGO, BUT RETAINED MY OVARIES. IF I TAKE HORMONE THERAPY IN MENOPAUSE, DO I NEED TO TAKE BOTH ESTROGEN AND PROGESTIN?

Modern medical thought is that taking estrogen and progestin together protects us against endometrial cancer. If you have no uterus, current medical practice indicates that you need only estrogen replacement therapy. That appears to be one plus of having had a hysterectomy: There is no need to take progestin, a synthetic progesterone, which has its own set of side effects.

36. I HAD A TOTAL HYSTERECTOMY EIGHT YEARS AGO AND I'VE BEEN ON ESTROGEN EVER SINCE. WHY DO I NOW HAVE HOT FLASHES AND NIGHT SWEATS?

This is a medication and dosage question that you should discuss with your physician. Perhaps whatever other bodily sources of hormones you might have had (such as conversion of some male hormone to estrogen) that added to your body's estrogen levels have now run out, and you need a higher dose or a different preparation.

37. WHY DO SO MANY WOMEN IN THE UNITED STATES AND CANADA UNDERGO HYSTERECTOMIES?

That is a hard question to answer. According to the Centers for Disease Control, the most common reason for performing a hysterectomy is uterine fibroids. The other most common diagnoses for which hysterectomies are performed are endometriosis, prolapse, cancer, and endometrial hyperplasia. In the United States, about one of every three women has a hysterectomy by the time she is sixty.

38. WILL MY DOCTOR BE OFFENDED IF I ASK TOO MANY QUESTIONS OR ASK FOR A SECOND OPINION BEFORE I AGREE TO HAVING A HYSTERECTOMY?

An interested physician should welcome your interest and participation in the decision. Many physicians and surgeons have told me that a knowledgeable, well-informed patient is easier to care for and does better in recovery. If you want to seek a second opinion, your surgeon should assist you in locating another surgeon to review your case. Or, you may choose to find your own. Under many insurance policies, a second opinion for surgery is required. Unless you must undergo emergency surgery, you have time to learn your options and to think through your plans. It is always wise to be an educated consumer, particularly of medical care.

I believe that there is no such thing as minor surgery. I agree wholeheartedly with Carol Ann Rinzler's definition of surgery in her book, *Feed a Cold, Starve a Fever: A Dictionary of Medical Folklore:*

operations (surgical)
    "It's just a minor operation." Although some surgical procedures are clearly less complicated than others, there is no such thing as a minor operation when general anesthesia is involved. Anesthesia al-

ways carries the possibility of complications, up to and including cardiac arrest and brain damage.

Need convincing? Then consider this: statistics compiled by the Commission on Professional and Hospital Activities of the American College of Surgeons show that in 1986, the last year for which these figures were available, 543 people died after appendectomy, 119 after tonsillectomy.

From *Feed a Cold, Starve a Fever*. Copyright © 1991 by Carol Ann Rinzler. Reprinted with permission of Facts On File, Inc., New York.

CHAPTER 5

What Is Hormone Replacement Therapy and How Does It Work?

The pros and cons of hormone replacement therapy (HRT) continue to bring the largest number of questions from women who attend the programs at which I speak. The second-largest number of questions concern cancer that may result from or be enhanced by estrogen. Although newspapers, magazines, television, and radio have covered the estrogen/cancer relationship frequently, straightforward answers seem to be in short supply. Headlines can range from DOUBT REMAINS ABOUT ESTROGEN FOR MENOPAUSE (*The Wall Street Journal*, September 12, 1991) and WOMEN FACE DILEMMA OVER ESTROGEN THERAPY (*The New York Times*, September 17, 1991) to headlines hailing estrogen therapy for its role in preventing osteoporosis and for lowering by one half women's death rate from coronary heart disease.

The controversy is not new—nor is it over—but the information is improving. The debate goes back at least to 1965, when a book was published that was touted as promising that taking estrogen equaled drinking from the fountain of youth. The best-selling *Feminine Forever* by Dr. Robert A. Wilson sent large numbers of women to their physicians for estrogen. The number of women using it escalated over the course of the next decade. Then, in December 1976, the prestigious *New England Journal of Medicine* published two articles which demonstrated a strong association between estrogen replacement therapy and the development of endometrial (uterine) cancer. Suddenly, it seemed, the fountain of youth had run dry.

In the early 1980s physicians began to prescribe estrogen with progestin to be taken in just the way your body would produce those two female hormones if your ovaries were still functioning. Physicians believed that women whose uteri were intact could, if both hormones were taken, thrive on the benefits of estrogen while being protected from endometrial cancer, which appeared to have occurred as the result of taking estrogen alone. Even with this more conservative approach, physicians and women were much more cautious about hormone replacement therapy than they were in the 1960s. Today, fewer than 20 percent of eligible women use hormone replacement therapy, but that number is growing.

It is growing because today's women are educated, interested, and accustomed to getting their questions answered. They are showing up at health education programs and in physicians' offices in ever-increasing numbers. They want to know about these midlife physiological and psychological changes and what they can do about them.

As women demand more answers, the impetus for more research into women's health issues has escalated steadily since the fall of 1989. In 1991, following her appointment as the first female director of the National Institutes of Health (NIH), Dr. Bernardine Healy initiated a request to Congress for five hundred million dollars for women's health research. As a result, an NIH Office on Research on Women's Health has been established. Perhaps, at long last, definitive answers to the estrogen replacement therapy risk/benefit questions will be forthcoming.

An article and editorial in the *New England Journal of Medicine* (September 12, 1991) underscores the current healthy research climate. The article, "Postmenopausal Estrogen Therapy and Cardiovascular Disease," a ten-year follow-up from the Nurses' Health Study, strongly suggests the cardioprotective effect of ERT in older women, while the editorial calls for ". . . action, not debate." The Nurses' Health Study was begun in 1976 and conducted at Harvard by Doctors Meir J. Stampfer,

Graham A. Colditz, Walter C. Willett, JoAnn E. Manson, Bernard Rosner, Frank E. Speizer, and Charles H. Hennekens. This study suggests that ERT can cut the risk of heart attacks in women in the first ten years after menopause by 40 to 50 percent. Other studies continue to demonstrate that estrogen therapy can prevent the osteoporosis that plagues women after menopause.

There are so many consumer questions about hormone therapy that, for purposes of easy reference, I have divided them into five separate chapters. This chapter will cover the questions from women who either take estrogen now or want to or plan to take it. Chapter 6 deals with questions from women who want to know if they can take estrogen despite the fact that they have certain other medical conditions. Chapter 7 includes questions about cancer and ERT. Chapter 8 looks at ERT and osteoporosis, and Chapter 9 answers questions concerning ERT and heart disease. Chapter 10 covers questions from women who seek information about non-hormonal therapy, because they cannot or do not wish to take ERT.

39. WHAT IS ESTROGEN AND WHERE DOES IT COME FROM?

Estrogen is one of the female hormones produced naturally by the ovary. It is the hormone that turns us from girls to women at puberty, making most of us fertile and able to have babies until menopause. The estrogen used in hormone replacement therapy may come from natural or synthetic sources. Your physician will usually determine which form of estrogen is best for you.

40. HOW IS ESTROGEN REPLACEMENT GIVEN?

Estrogen replacement therapy may be given by patch, pill, cream, or injection. There are advantages and disadvantages to each method of administration.

The transdermal patch is the newest form of ERT. Transdermal literally means "through the skin," and that is how the patch works. The patch is approximately the size of a silver dollar and it is transparent. It is usually worn on various locations on the abdomen or buttocks and delivers estrogen literally through the skin in a slow, even dose just as your ovary would if it were still functioning. There are two dosage sizes: The larger dose patch is slightly larger in size and is oval in shape. Regardless of the dose, each patch lasts three and a half days, so you change to a new patch twice a week.

The pill is obviously taken by mouth and comes in different dosages and preparations. Estrogen cream is used vaginally. Although it is absorbed through the bloodstream and thus goes elsewhere in the body, it is used primarily for menopausal problems limited to the vaginal or urinary tracts, such as vaginal dryness or urinary tract infections. Injectables have been around for a long time. These estrogen shots are given once a month by a physician. Other new methods of ERT delivery are being developed, and no doubt will lengthen this list. There is talk of a new subcutaneous pellet form and of a new topical cream, as well as an estrogen/progestin combination patch.

41. WHAT IS THE DIFFERENCE BETWEEN BIRTH CONTROL PILLS AND ERT?

Birth control pills, even the newer low-dose pills, contain a much larger amount of estrogen than the doses prescribed to counteract postmenopausal symptoms.

42. DOES ESTROGEN REPLACEMENT THERAPY DECREASE A WOMAN'S RISK OF DEVELOPING CARDIOVASCULAR PROBLEMS?

The cardiovascular issue is a very complex one as far as ERT is concerned. At this time, there are no long-term, clinical trial results demonstrating cardiovascular-protective results from ERT in general although some studies suggest that effect. In addition, no estrogen product has an indication from the Food and Drug Administration (FDA) stating that estrogen products have any cardiovascular benefit. This is an area for further research, and I hope that more and larger studies will provide some answers to this important question. As I mentioned earlier in this chapter, the follow-up report from the recent study by Harvard medical researchers, in which more than one hundred twenty thousand female nurse were surveyed by mail since 1976, showed the incidence of heart attacks in the postmenopausal years to be much lower among the nurses who took estrogen postmenopausally than among those who never took estrogen.

There are calls for larger-scale heart disease studies to be undertaken, such as those that have already been done with regard to men's cardiovascular health, to determine the final answer to this question. What is known concretely today is that for women, death from heart disease is eight times greater than death from breast cancer, which afflicts one in nine American women. (More about breast cancer in Chapter 7. More about coronary heart disease in Chapter 9.)

43. WHAT HAPPENS WHEN YOU STOP TAKING ERT?
Either all or some of your symptoms will return, or you will have passed through the period of time during which overt symptoms will bother you. However, the protective and nourishing effect of ERT on the many organs and processes of your body will also cease.

44. IS THERE A TIME AFTER WHICH THE PARTS OF MY BODY—HEART, BONES, BLADDER, VAGINA—NO LONGER NEED ERT?
There are a number of opinions on this point. Some experts believe that you should take the smallest amount of ERT (or estrogen and progestin therapy, HRT, if you have a uterus) that you can take that still alleviates your symptoms, and that you should take it for the shortest amount of time. If you are taking it only to get rid of certain menopausal symptoms, such as hot flashes and night sweats, that probably means taking it for approximately two years. However, in order for ERT to provide

effective osteoporosis protection, you should probably continue ERT until the process of bone loss begins naturally to slow down, which is usually around age sixty-five. (Osteoporosis questions are answered in Chapter 8.) Regarding the question of whether or not ERT offers protection from heart disease, the jury is still out, although the verdict as of this writing seems to point to continuing therapy. The question of when to quit should be discussed thoroughly with your physician. *Do not ever alter or stop taking ERT without your physician's knowledge and approval.*

45. I AM ON HORMONE THERAPY. HOW OFTEN SHOULD I SEE MY PHYSICIAN?

Hormones are powerful drugs. When you are on therapy, you have a responsibility to see your physician regularly, at least annually. Many physicians with whom I discussed this point prefer to see their ERT and HRT patients twice each year and, certainly, to be contacted if any changes occur between visits. You need to manage your own midlife health care in partnership with your physician. Don't be a no-show!

46. CAN A MENOPAUSAL WOMAN ON HORMONE REPLACEMENT THERAPY BECOME PREGNANT?

Usually not, because the ovaries are no longer delivering eggs to be fertilized. However, gynecologists caution you to use contraception for at least a year after your last period, just in case your ovaries decide to turn on one last time. Surprise!

47. WILL ESTROGEN HELP DEPRESSION?
It may, *if* it is a minor depression caused by the effect of your menopausal symptoms, particularly lack of solid sleep. A deep discussion with your physician is warranted on this issue. If your doctor believes that estrogen will help and prescribes it, you should start to feel better and to sleep better in a remarkably short period of time, perhaps a week or two.

48. I TAKE ESTROGEN PILLS. SHOULD I TAKE THEM AT THE SAME TIME EACH DAY?
Becoming a creature of habit with medication is a very good idea for several reasons. First of all, your body will become accustomed to how the pills are spaced and your hormone levels on average will be kept more constant. Even more important: Taking pills at the same time each day assures that you will remember to take them. If you are using the patch, it's a little easier because there is a place right on the cardboard cover of the box (which contains a one-month supply) to mark the days when you change patches. Women use estrogen creams in several ways—some on an as-needed basis and others daily—so in such cases, you could mark your appointment calendar to indicate how you and your physician have set up your own administration system. Of course, injections are

given at your physician's office. Doctors indicate that these shots should be pretty evenly spaced.

At a program in Pittsburgh, a woman told us of her failproof dosage system: She puts her pill container next to her toothbrush, and she has been brushing her teeth first thing each morning for more years than she cared to discuss. She has never forgotten to take her pill. Another woman changes her patch on Fridays when she goes to the beauty parlor and on Mondays before she begins her work week. Whatever habit or reminder triggers the right response for you is a good medication system.

49. IF I TAKE ESTROGEN AND I'M IN THE SUN A LOT, WHAT WILL HAPPEN?

Many women on ERT have been dismayed to find that patches of colored pigment appear on their faces if they spend time in the sun. Moreover, the incidence of skin cancer is on the rise whether or not you take estrogen. Many years ago, ladies wore large brimmed hats and carried parasols or sun umbrellas to protect themselves from the sun's rays. Inasmuch as the porcelain complexion is again in vogue, this time for both health and fashion reasons, these are not such old-hat ideas! Being in the sun a lot without protecting your skin with a sun block of at least an SPF 15, regularly reapplied, is passé. So cover up with lotion, sun hats, and sunglasses.

50. I SWIM, SNORKEL, AND SCUBA DIVE. CAN I USE THE TRANSDERMAL PATCH IN THE WATER?

The patch is held on by a ring of adhesive and usually stays in place no matter what your water activity. The only complaint we heard was from one woman who regularly uses a Jacuzzi. Her answer to a loosening patch caused by the jets of bubbling water is to remove it and place it on its original backing until she comes out of the tub and then she puts it right back on.

51. WHAT CAN BE DONE TO PREVENT SKIN IRRITATION FROM PATCH THERAPY?

Women have complained of two separate problems. One problem relates to their sensitivity to the adhesive material; the other (an infrequent complaint) is a sensitivity to the estrogen itself. In most cases, both of these problems can be solved by placing the patch on a new spot on your abdomen or buttocks each time you change patches. Originally, I showed a slight sensitivity to the estrogen in the form of a slight pink mark on my skin, but with repeated use it just stopped occurring. Women whose skin continues to show sensitivity to the adhesive may need to switch to another form of ERT.

52. ARE THERE ANY NEGATIVES TO ERT?

Estrogen replacement therapy can be very effective, but it should always be considered very carefully. A small percentage of women experience side effects such as nausea, fluid retention, swollen breasts, weight gain, and even vaginal discharge. In women with an intact uterus who use ERT alone, cancer of the lining of the uterus (endometrium) has been found to occur more frequently, although adding progestin for twelve days each month appears to nullify this cancer risk.

The United States Department of Health and Human Services booklet, *The Menopause Time of Life*, lists women who should not use ERT as those having heart disease, endometrial cancer, breast cancer, stroke, migraine headache, high blood pressure, blood clots, and certain other circulatory disorders. It also cautions those women with a family history of cancer to be extra careful. (More cancer questions are considered in Chapter 7.) The Health and Human Services booklet, published in 1986, also urges cautious consideration of ERT for women with liver disease, gallbladder disease, and those with diabetes. (You may obtain a copy of the booklet by writing to the National Institute on Aging, NIA Information Center, 2209 Distribution Circle, Silver Spring, MD, 20910.) However, since the advent of the transdermal patch in 1986, which allows the hormone to initially bypass the liver, women who wish to consider ERT but who were unable to take it because of certain kinds of hypertension, clotting problems, gallbladder disease, and liver disease, should rediscuss it with their physicians. They may be able to use the patch.

53. How do I know whether I can take ERT?
This again is an evaluation to be made by your physician. Current medical thinking seems to indicate that you cannot consider ERT if you have a personal or a family history of uterine cancer or estrogen-dependent breast cancer. Chapter 7 discusses this questions more fully.

54. Does every woman need ERT?
Physicians indicate that up to one third of women do not need ERT. Some menopausal women seem able to make a small amount of estrogen or to convert other hormones to estrogen or to be able to use estrogen that is stored in fatty tissues, compensating to some degree for the ovaries' cessation of estrogen production. Often these are the women who sail through menopause. Women whose symptoms are uncomfortable and who have no medical reason not to take estrogen often choose to take ERT. Some women simply cannot take it for medical reasons. The four out of ten postmenopausal women who ultimately will have osteoporosis probably need to discover that they are at high risk as early in their lives as possible. Their physicians will probably encourage them to include calcium-rich foods in their diets, calcium supplementation, plenty of approved exercise, to quit smoking, and to take ERT to reduce their risk. (See Chapter 8 for more on osteoporosis and ERT.)

In terms of protection against coronary heart disease, the indicators seem to support the effectiveness of ERT, but conclusive evidence is still unavailable and large-scale studies of women and heart disease desperately need to be conducted. A 1991 article in the prestigious British medical journal *Lancet* underscored estrogen's dramatic effect on serum cholesterol levels. Women taking it showed a significant increase in HDL, the "good" cholesterol, and a decrease in LDL, the "bad" cholesterol. This is important because this improvement in serum cholesterol levels decreases arterial plaque and thus decreases the obstruction to blood flow in the arteries.

55. CAN I TAKE HRT IF I HAVE A BENIGN FIBROID TUMOR IN MY BREAST OR UTERUS?

Many women at the programs asked this question. In the past, the presence of fibroid tumors in a woman's breasts or uterus has often been the reason why her physician steered her away from estrogen. The low dose of estrogen used today in replacement therapy has changed the answer to this question from a definite *no* to a perhaps. Discuss it anew with your physician.

56. EVERYONE KEEPS TALKING ABOUT THE PATCH BYPASSING THE LIVER AS A GOOD THING. WHY IS THAT?

When estrogen passes through the digestive system (as it would if ingested in pill form), the liver extensively metabolizes it, perhaps changing it in some important ways. Current medical theory is that by allowing estrogen to initially bypass the liver, the patch system of estrogen replacement therapy may permit you to use ERT whether or not you have certain other medical conditions such as liver or gallbladder disease, certain forms of hypertension, or other circulatory system problems. Moreover, because the estrogen is not changed in the liver, your dosage is more easily monitored by your physician.

57. WHAT CAN BE DONE IF YOU ARE ALREADY ON ERT AND YET ALL YOUR UNCOMFORTABLE SYMPTOMS DO NOT DISAPPEAR?

Sometimes changing the kind of estrogen you are taking or increasing the dose will help, but this is not something you can consider without seeing your physician. In the case of hot flashes, many women, including me, have had hot flashes disappear by taking Vitamin E along with ERT. Although there is no scientific proof that Vitamin E alleviates hot flashes, it does appear to help some of us. During the period when I experienced hot flashes and thought they were just sweats induced by stress, I took four hundred milligrams twice a day and the flashes went away. Ask your doctor if you can try Vitamin E.

58. WHAT CAN I DO ABOUT THE AMOUNT OF WEIGHT I'VE GAINED SINCE BEGINNING ERT?

Estrogen does not increase your weight, but some women insist that it can make you hungrier. About one in four women report a slight weight gain after going on ERT. Whether this is due to water retention or to an actual increase in fatty tissue is not clear, but it still shows on the scale and in how our clothes fit. To add to the problem, our need for estrogen at menopause comes at a time when our metabolism is slowing down, and we may gain weight for that reason as well. More on weight gain, metabolism, and water retention in the chapters on lifestyle, Chapters 13 and 14.

59. DOES HORMONE REPLACEMENT THERAPY CAUSE MENSTRUAL PERIODS TO CONTINUE? FOR HOW LONG? WILL I GO ON MENSTRUATING FOREVER?

"I thought one of the benefits of menopause was having no more periods!" This appears to be the number-one complaint about HRT from women who have not had hysterectomies. Yes, HRT—the cycling of estrogen with progestin, just as your body did before menopause—will probably bring back your periods. It also may help protect you against endometrial cancer. How much and when you bleed on HRT is a subject that you and your doctor must discuss. On average, your postmenopausal periods start the day following your last progestin tablet. If bleeding occurs at any other time, it is important that you contact your physician. No, periods are not forever even if it seems that way. They usually dwindle to nothing after a few years of HRT.

60. MY PERIODS ARE LESS AND LESS PREDICTABLE. CAN I START TO TAKE HRT CLOSE TO THE END OF MY PRE-MENOPAUSE AND AVOID THE ONSET OF SYMPTOMS ALTO-GETHER?

Not usually. Your ovaries are still producing estrogen even though your periods may be episodic. Therapy might give you way too much estrogen, causing build-up of the lining of the uterus and possibly creating a precancerous condition. The premenopause can last from one year to even five years or longer. It isn't considered to be over until you have gone a full year without a period. Most physicians say that you should not start on therapy until twelve months without a period have passed. However, I suggest you discuss this point with your own physician. Your medical history may enable you to use it sooner.

Life-style changes, with or without hormone replacement therapy, are often required in order to enhance the quality, not to mention the longevity, of our lives. We might consider paralleling our personal view with that stated in the International Plan of Action on Aging formulated at the United Nations meeting in Vienna: Diseases do not need to be essential components of aging.

CHAPTER 6

❧

Am I a Good Candidate for Estrogen Replacement Therapy?

According to the questionnaires that women voluntarily fill out and leave with us after the programs, an average of 70.2 percent wanted to know more about ERT. Women indicated that they wanted more knowledge about the psychological effects of menopause as well as the physical effects of menopause (64.2 percent and 62.4 percent, respectively). They sought more information concerning weight gain and anxiety/irritability (59.5 percent and 58.3 percent). They felt they needed to know more about insomnia (50.4 percent), mood swings/depression (46 percent), vaginal dryness (45.2 percent), night sweats (42.8 percent), hot flashes (40.2 percent), and, last of those questions asked, sexual difficulties (38.9 percent). The percentages represent averages of the ten cities studied and yet also represent the information gleaned from women in any given city, showing little geographical differences in the order of the symptoms that women felt they needed to know more about.

What is not asked in the questionnaire, but is determined from the index card questions that the audience members pass forward for the panelists to answer, are specific questions about individual contraindications, or reasons against using ERT. Many of those questions are from women who want to know whether they can take estrogen if they have had breast or other kinds of cancer, or if there is a history of cancer in their families. Questions concerning cancer and ERT will be addressed in Chapter 7.

There are a number of other physical conditions that are of concern to women who are considering ERT. A delicately boned, smartly dressed woman approached the podium in Seattle and spoke to the physician panelist who had discussed ERT in his presentation. She said that she had a history of blood clots in her legs, but that her menopausal symptoms had gotten so bad that her physician had agreed to short-term ERT. Even this short-term dosage was quickly stopped, however, when another blood clot developed. She wanted to know whether she could chance ERT again, because her menopausal symptoms were making her miserable. The doctor reminded her of the chart he had shown during his talk that indicated that ERT was not generally recommended for women with circulatory or vascular problems, but that some form of ERT might be able to be considered, depending on the patient's medical history and problem. Then he explained to her that only she and her physician could work out the solution to her individual problem.

Many such personal questions are asked at the programs. Women need to take questions about estrogen replacement therapy and their personal or family history of phlebitis, varicose veins, stroke, uterine fibroids, fibrocystic breast disease, gallbladder disease, liver disease, and endometriosis to their own doctors, who are familiar with their unique medical histories and who can make decisions with the patient based on these pertinent facts. What I can do in this chapter is offer the current scientific thinking in regard to these questions. Remember, however: You are an individual with an individual problem. Only your own doctor can answer your particular questions.

61. I PASSED THROUGH MENOPAUSE WITH NO SYMPTOMS THAT I KNOW OF. DO I NEED ERT?
Maybe not. Physicians indicate that up to one third of postmenopausal women do not need ERT. Some menopausal women are able to continue to produce enough estrogen themselves to ward off symptoms and to prevent the metabolic changes that may lead to postmenopausal osteoporosis and, perhaps, to heart disease. The estrogen source for these women may be the adrenal gland, which

is located above the kidneys, which may convert its hormone, androstenedione, to estrone (estrogen). This conversion takes place in the fat cells, so overweight women may produce enough estrogen to compensate for their ovaries' cessation of estrogen production. These are often the women who sail through menopause. However, these asymptomatic women should also be evaluated by their physicians to make sure that their hormones are balanced. A new study suggests that they need to make sure that they have not become candidates for cancer of the uterus, a situation that may result when estrogen is present without progesterone in women who have an intact uterus. See your doctor!

62. WHAT CAN ERT DO FOR ME?

ERT or HRT (if you have an intact uterus) can resupply your body with the female sex hormones that will positively affect hundreds of different processes in your body. Estrogen can alleviate menopausal symptoms such as hot flashes, night sweats, vaginal dryness, and other vaginal and pelvic changes; it can prevent postmenopausal osteoporosis; it can offer relief from insomnia; may help with memory loss; may return your desire for sex; may return some of your skin's elasticity; and it may stop the palpitations, mood swings, and the feelings of anxiety and minor depression that may be related to menopausal symptoms. Further, if current research into the cardio-protective effect of estrogen proves this beneficial effect, estrogen may protect you against heart disease and heart attack. That's a pretty compelling list.

63. WHO IS NOT USUALLY CONSIDERED A GOOD CANDIDATE FOR ERT AND HRT?

As I mentioned earlier in this chapter, ERT is not usually indicated for any women who have heart disease or who have had endometrial or breast cancer or any other kind of cancer that is stimulated by estrogen, or any women with a family history of these types of cancer. This will be covered more fully in Chapter 7. Women with chronic or acute liver disease or with genital bleeding that is abnormal and unexplained also are not considered to be good candidates for hormone replacement therapy. Caution is also indicated for women with gallbladder disease and those with diabetes.

64. I HAVE UTERINE FIBROIDS. IS IT SAFE FOR ME TO TAKE ESTROGEN?

Please refer to question 55 on page 88 for a general answer to the issue of fibroids and ERT. This is another of those situations that I mentioned earlier in which you have to work out the answer with your doctor.

65. I'VE HAD PHLEBITIS, SO I'M WORRIED ABOUT TAKING ERT, BUT I DO NEED RELIEF FROM MENOPAUSAL SYMPTOMS. CAN I TAKE ESTROGEN?

Scientific opinion is divided on this point. Phlebitis is a circulatory problem in which a vein is inflamed. Physicians with whom I've discussed this subject believe that some form of ERT may be attempted if it is carefully monitored. Again, you and your doctor are the only ones who can answer that question after you weigh the risk-to-benefit ratio. If the answer is "no," you may find some of the nonhormonal therapies discussed in Chapter 10 to be of help.

66. I HAVE VARICOSE VEINS IN BOTH OF MY LEGS. CAN I TAKE ESTROGEN?

Varicose veins do not make ERT impossible to take, but because they indicate a circulatory problem, modern medical thinking suggests that prescribing it requires thoughtful and deliberate consideration. Your physician can help you decide after he or she reviews your medical history, completes a thorough physical examination of you, and considers the risks versus the benefits.

67. I HAVE ENDOMETRIOSIS AND WAS UNABLE TO HAVE CHILDREN. NOW I AM SUFFERING TERRIBLY WITH MENO-PAUSAL SYMPTOMS. IS ERT OUT OF THE QUESTION FOR ME? MY FORMER PHYSICIAN SAYS "YES," BUT SOME OF THE BOOKS I READ SAY I MIGHT BE ABLE TO TAKE IT FOR A WHILE TO RELIEVE THE WORST OF THE SYMPTOMS. IS THAT POSSIBLE?

Endometriosis is a condition in which the tissue that normally is found inside the uterus is displaced and starts to grow elsewhere in the pelvic cavity. Let's review estrogen's job in the body, which is to enhance, or thicken, the lining of the uterus, the endometrium. Progesterone's job is to precipitate the shedding of that lining. Endometriosis is wayward tissue, but still is an estrogen-dependent condition. It may be arrested today with laser treatments and is also expected to clear up if the ovaries are surgically removed or when, at menopause, they no longer produce estrogen. So, you can see why estrogen replacement therapy would seem to defeat the clearing-out of the wayward tissue. Since your question indicates that you have found or are in the process of locating a new gynecologist, I suggest that you discuss the subject of endometriosis and hormone replacement therapy with your doctor. You must decide, in cooperation with your physician, what is best for you.

68. DOES THE FACT THAT I HAVE DIABETES PREVENT ME FROM TAKING ESTROGEN?

Today, because ERT is generally of such a low dose, it may be considered for women with diabetes, because estrogen seldom affects the metabolism of sugar. However, if your physician does put you on ERT, he or she undoubtedly will want to monitor your blood sugar carefully.

69. CAN SOMEONE WITH HIGH BLOOD PRESSURE (HYPERTENSION) CONSIDER ERT?

Estrogen taken by vaginal cream or by transdermal patch does not appear to affect blood pressure, because it does not cause the release of enzymes from the kidneys which may, in turn, raise blood pressure. That is because estrogen delivered transdermally (through the skin) does not initially go through the liver, the kidney, or the digestive system. In about five women out of a hundred this release of the enzymes (renin and angiotensin) from the kidney may occur with oral ERT, which may be contraindicated for someone with high blood pressure. If oral ERT is not for you, your blood pressure will increase shortly after you start taking the oral estrogens. Most physicians agree that women with hypertension can take some form of estrogen that will not usually alter blood pressure.

70. WILL ERT STOP ALL MY MENOPAUSE SYMPTOMS? Most physicians say that many women say "good-bye" to hot flashes, night sweats, uncomfortable sex because of changes in the vagina, and to most of the other unpleasant symptoms of menopause within a couple of weeks of beginning ERT. If your preparation and dosage of ERT leave you with some undesirable menopause symptoms, contact your physician. Maybe you can work out another dosage or preparation that will be of greater help to you.

For too long, we have made the mistake of making deities out of our doctors—a role the physician did not choose and does not want. As you have seen, this chapter indicates how great a role you must play when you are considering ERT or hormone replacement therapy with estrogen and progestin. Each woman needs to educate herself thoroughly about the medical conditions she has and how ERT may possibly have an effect upon them. Then, in concert with a trusting and comfortable patient/physician *partnership*, you can take control of your own decisions regarding your good health.

To me, Betty Friedan stated the imperatives of this kind of relationship in her introduction to the twentieth-anniversary edition of her landmark book, *The Feminine Mystique:*

> *And as women take control of their bodies, their selves, even as patients, male psychiatrists, obstetricians and gyne-cologists are being forced off their godlike pedestals to treat women and other patients as people . . .*

CHAPTER 7

Does Estrogen Therapy Cause Cancer?

Many women who choose not to take estrogen make that decision based on their fear of cancer. The Gallup survey referred to earlier revealed that in response to the following question: "To the best of your knowledge, what is the leading cause of disease-related death for women over 50?" The most-frequent answer (33 percent) given by the women surveyed was breast cancer, with heart disease running a close second (31 percent). However, heart disease is actually number one, killing eight times more women than breast cancer each year, yet it appears that it is breast cancer that we most fear. (Heart disease will be discussed in Chapter 9.) The relationship between cancer and estrogen therapy needs to be carefully examined before we attempt to answer the specific questions about cancer that are asked at the programs.

A slim, free publication from the National Cancer Institute, *What You Need to Know About Breast Cancer* (for your copy, write to the National Cancer Institute, 9000 Rockville Pike, Building 31, Room 10A24, Bethesda, MD, 20892), describes clearly what cancer is and how it grows. It explains that there are one hundred types of cancer currently known, among which are several kinds of breast cancer. The characteristic that is common to all cancer diseases is that abnormal cells proliferate and destroy tissues in the body.

The American Cancer Society indicates that in 1991, approximately 175,000 American women got breast cancer and that 44,500 died as a result of it. It is easy to understand why these facts strike fear in the minds of women. Yet there is more that we need to know. Let me cite a perfect example of the effect of educating yourself about the real facts of the cancer/estrogen conflict.

One woman caught me in the hotel lobby after a program. "I wonder if I could talk to you for a few moments. I have such a need to tell you how I feel about this estrogen and breast cancer situation. I know you're not a physician," she said, "you made that perfectly clear during your presentation. But please let me tell you how I feel, because I think it may help other women and, besides, I really can't go home and sleep tonight without getting this off my chest. No pun intended," she laughed.

It is these kinds of conversations that let me know what women actually glean from the information presented at the programs and how they really feel about the changes that occur in midlife. "I'm happy and eager to listen," I assured her as we sat down together.

"My grandmother died of breast cancer," she began. "She was eighty-nine, had been overweight since she was in her sixties, and she was sedentary. My mother died at seventy. She had breast and bone cancer and a stress-filled life. From the time she was my age, fifty, she was overweight and she didn't exercise. Mom's sister, my aunt, had the same general body and life-style profile, only she had eight kids. She also died of cancer that was first found in her breast.

"What a family history I have! But I figured I would escape cancer, because I was different. I'm a health-food nut and an exercise devotee, so I figured I could protect myself from following in the footsteps of the females in my family. That made me feel secure until my cousin, slim and in her forties, found a lump in her breast, which when biopsied was found to be malignant. She had a lumpectomy. Since then, getting breast cancer is always in the back of my mind. So I've been getting ready for menopause with a real chip on my shoulder.

"I've already had a small noncancerous growth removed from my left breast and I've read everything cautioning someone like me, with a personal medical history and a family medical history like mine, against taking ERT. I felt pretty discriminated against and hostile, especially now when news reports of medical studies keep informing me that there are many diseases that estrogen will, or may, prevent as women age. Big deal, I thought.

"I read everything I can get my hands on to try to understand the risks of breast and endometrial cancer for women who take hormone replacement therapy and I have felt cheated. Tonight turned that around for me. Now I know that if my menopausal symptoms are difficult for me and I'm given a choice, I can chose to take estrogen under careful medical supervision. I know now that I might be willing to take that gamble. Hallelujah!"

"Whoa," I said. "Before we go any further, tell me what brought you to that conclusion." I wanted to make sure that she had accurately heard what was said at our program and that she was reacting to sound medical information.

She continued, "The doctor who was discussing cancer said that although there is some evidence that taking estrogen may increase the risk of developing breast cancer, it is a small risk compared to the fact that estrogen can help alleviate my symptoms, can prevent osteoporosis, and may even decrease the risk of heart disease. As I said, I'm getting ready to gamble."

"You can only gamble if your physician agrees and if

you're willing to take the responsibility for seeing him or her at least twice a year, for having a mammogram once a year, and for learning and practicing breast self-examination once a month," I cautioned.

"I know that. Tonight's information really made me feel great. It made me feel like I could take control of what's happening to me. I just needed to tell you that now, instead of feeling like a victim, I feel empowered to make choices for myself."

How could a single lecture turn this intelligent and articulate woman's perceptions around so completely? It was simple. The program gave her the opportunity to learn the facts and to ask questions, and then to make a decision based on sound information.

It's no wonder that there is so much confusion about ERT and breast cancer. An article in *Consumer Reports* (September 1991) stated that the actual data concerning the relationship between breast cancer and estrogen therapy are not definitive. Conflicting results are produced by important studies. A major survey, in which researchers at Vanderbilt School of Medicine analyzed twenty-eight of the ERT studies published over the last fifteen years, found no increased risk for breast cancer at all. These researchers concluded that "the combined results from multiple studies provide strong evidence that hormone therapy (at the current standard dosage) does not increase breast cancer risk."

Compare this finding to The Nurses Cohort Study at

Harvard, reported in the *Journal of the American Medical Association* (September 28, 1991). This ongoing prospective study (a study designed to gather data over a specified period of years) in which more than one hundred twenty thousand women reported, found that women who used estrogen in the past, but who no longer use it, return to the same risk of breast cancer as women who have never used estrogen. No increase in breast cancer was found in women who stopped ERT, even after taking it for ten years. But, the study found, a woman who is currently on ERT has approximately a 35 percent higher risk of developing breast cancer than a woman who does not use it, and that remains true no matter what dose of estrogen she is taking or for how long she has taken it. Further, a woman has a greater risk of getting breast cancer if her mother or sister have had it— in which case the risk is doubled. Additionally, these investigators indicate that estrogen may not actually cause cancer, but may cause the growth of an existing tumor. This could actually be helpful news, because it could mean that an undetected tumor would not stay hidden and could be treated at an early stage.

Most of the doctors surveyed are inclined to agree that if there is an increased risk of breast cancer with ERT, it is very, very small, perhaps from one to one and a third in one thousand women per year. Yet breast cancer affects one in every nine women in the United States over their lifetimes. According to Dr. Morris Notelevitz's article in *Menopause Management*, a professional medical journal, "While the incidence of breast cancer does not appear to be menopause-related, it is known to increase with age."

Clearly there is a need for medical guidelines in determining the safety of hormone therapy in postmenopausal patients with breast cancer or with a family history of breast cancer. Two of the potential major risk factors for ERT are breast and endometrial (uterine) cancer. Women at the educational programs submit an extremely large number of question-cards about these concerns, but their questions basically fall into the ten areas covered by the questions in this chapter.

71. IS THERE A RELATIONSHIP BETWEEN HORMONE REPLACEMENT THERAPY AND CANCER?

Some studies indicate that *long-term* ERT increases the risk of breast cancer, but only slightly. There also appears to be a relationship between estrogen and cancer of the uterus. The fear of breast cancer is the number one reason why women choose not to take estrogen today, and the fear of uterine cancer was the reason why most estrogen replacement therapy was withdrawn in the 1970s. It is important that women understand these relationships and put them into proper perspective.

In the National Family Opinion Survey, 73 percent of the one thousand women surveyed believed that the risk of breast cancer was one disadvantage of ERT; 58

percent thought that the risk of uterine cancer was a disadvantage of therapy. Although breast cancer accounts for 27 percent of all cancers in women, its number one position was overtaken by lung cancer in 1989. As I mentioned earlier, even though breast cancer is what most women fear, eight times more women will die from coronary heart disease than from breast cancer. While endometrial cancer is not a concern for women who have had a hysterectomy, for women with an intact uterus, estrogen therapy does increase the incidence of cancer of the endometrium, the lining of the uterus. But that risk may be eliminated by taking progestin, the other female hormone, along with estrogen (and with careful monitoring by your doctor) so that the lining of the uterus is shed and therefore does not thicken. The risk of endometrial cancer is small and the cancer is easily detectable, so it can be found early and treated quickly.

72. WHAT ARE THE SYMPTOMS OF BREAST CANCER?

Breast cancer can cause many symptoms such as a lump or thickening in the breast or under your arm, a change in the shape or size of your breast, nipple discharge, or a change in the color or texture of the skin of the breast. You should see your doctor if you notice any change in your breasts. In the case of breast cancer, early diagnosis is very important, because the earlier breast cancer is found, the better the chances are for a complete recovery. It is important that we women practice monthly breast self-examination. Your doctor should be able to give you information on how to do this effectively.

73. IF I HAVE A FAMILY HISTORY OF BREAST CANCER, CAN ERT BE DANGEROUS?
Women in your situation are rightfully confused about the safety of ERT. Many such women do not want to take it and many physicians do not prescribe it. According to the National Institutes of Health, the risk of breast cancer doubles for a woman whose mother or sister has had the disease. If the mother's cancer began before menopause, her daughter's risk is slightly higher. Deciding to take estrogen replacement therapy in such situations often boils down to a quality of life issue—weighing uncomfortable or debilitating menopausal symptoms against these risks. With this kind of family history and in cases of debilitating menopausal symptoms some physicians may, occasionally, consider carefully monitored short-term estrogen therapy. Doses vary and a very low dose of estrogen may solve your problem.

74. I HAVE HAD BREAST CANCER. CAN I TAKE ERT?
The answer is usually no. If you have had cancer in one breast, your risk of getting it again is four to five times higher than for women who have not had breast cancer. National Institutes of Health studies indicate that long-term ERT increases the risk of breast cancer slightly.

Other studies do not confirm this, however. This is why you and your doctor must carefully analyze your personal risk. That analysis should include tests to determine whether your cancer was hormone-fed. An estrogen and progesterone receptor test can tell whether these hormones promoted the growth of the cancer. This information may help your doctor to decide whether hormone treatment is possible for you.

75. SUPPOSE YOU HAD BREAST CANCER AND CANNOT TAKE ERT. IS THERE AN ALTERNATIVE?

Chapter 10 details nonhormonal therapies for the symptoms of menopause. Many of these practices work very well for women who cannot or do not wish to take hormones.

76. WHAT IS THE NEW DRUG THAT IS UNDER CONSIDERATION FOR BLOCKING THE RECURRENCE OF BREAST CANCER OR PREVENTING IT IN HIGH-RISK WOMEN LIKE ME?

The National Cancer Institute has a new double-blind study under way of 16,000 high-risk women to see whether treatment with tamoxifen, a powerful drug, works to *prevent* breast cancer. Tamoxifen is believed to block cancer in the breast and to mimic the protective effects of estrogen elsewhere in the body. The results of the new study are five to eight years away. (In this double-blind study, half the women receive tamoxifen; the other half get a placebo.)

77. How often should I have a mammogram?
The American Cancer Society general guidelines recommend that women have a baseline mammogram between the ages of thirty-five and thirty-nine and then one every one to two years between ages forty and forty-nine. After the age of fifty, women should have a mammogram annually.

78. Does taking estrogen with progestin really protect me against endometrial cancer?
When estrogen and progestin are prescribed in concert, current scientific thinking is that your own monthly cycle is duplicated in a way that will protect you from endometrial cancer. The estrogen builds up the lining of the uterus and the progestin causes it to slough off, ridding your body of this build-up of endometrial tissue that lines the uterus. This is how the female hormones estrogen and progesterone produced by your ovaries work in your own body prior to menopause.

79. What is endometrial hyperplasia?
This is a condition caused by excessive buildup of the tissue (called the endometrium) lining the uterus. It is the most common cause of abnormal bleeding in women over forty years old. Treatment with progestins is usually

effective in alleviating this condition. If not treated, hyperplasia can develop into a precancerous condition. That is why any unusual bleeding should be reported to your doctor *at once*. If hyperplasia occurs, you need to see your doctor so that immediate steps can be taken to reverse this situation. It is the progestins taken in hormone replacement therapy that cause the endometrial lining to be shed in the form of withdrawal bleeding each month. Withdrawal bleeding is so named because it occurs after the progestin is withdrawn, or stopped. There is still much scientific study and discussion about the proper dosage of progestin required in HRT to allow sufficient shedding of the endometrial tissue while keeping bleeding to a minimum.

80. CAN I TAKE ESTROGEN WITHOUT PROGESTIN, IF I'VE HAD A HYSTERECTOMY (OVARIES INTACT)?

Current medical thinking is that if you have no uterus, you do not need to take progestin, since without a uterus (and thus without an endometrium) you cannot suffer from endometrial hyperplasia, nor is there risk of endometrial cancer as a result of taking estrogen alone.

81. DO I HAVE TO TAKE BOTH ESTROGEN AND PROGESTIN IF I HAVE MY UTERUS INTACT?

It is a well-accepted fact that in women with an intact uterus estrogen use alone (without progestin) raises the relative risk of endometrical cancer from about one case per thousand per year to between four and eight cases per thousand per year. Medical experts believe that the risk can return to one in one thousand with the addition of progestin approximately twelve days per month. (Yet the Food and Drug Administration has not actually made a decision concerning whether progestin cycled with estrogen should be prescribed as a preventive measure against endometrial cancer.)

82. I TAKE ESTROGEN ALL MONTH LONG. I TAKE PRO-GESTIN FOR THE FIRST TWELVE DAYS OF THE MONTH AND THEN I GET MY PERIOD. WILL I EVER STOP HAVING MY PERIODS?

The fact is that 80 percent of women on the combined therapy (HRT) will have periods. The good news about those periods is that you know the exact date when you're going to get them and you can plan accordingly. The better news is that the periods will usually diminish over time. So patience is a virtue in this case. Few women relish this aspect of HRT, *but* there is an important trade-off: greater peace of mind in possibly being better protected against endometrial cancer than if you were on ERT alone or if you stopped all therapy and suffered with hot flashes and other symptoms.

Today, many physicians believe that when the risks and benefits of ERT or HRT are weighed, the benefits far outbalance the risks. At a September 1990 news conference in New York, however, physicians conceded that the use of hormone treatment as preventive medicine for menopausal symptoms was still a matter of scientific debate.

Some scientists think that further studies may indicate that estrogen use of more than ten years' duration may bring with it increased risks of cancer. Yet I know women who rode out the chilling estrogen/cancer scare of the 1970s and who have now been on ERT for more than thirty years. They feel and look great. On the other hand, maybe they were just fortunate.

Obviously, there is a crying need for more and better studies as millions of women try to decide whether to undertake hormone therapy. By working in close concert with your doctor to weigh your own medical history against ERT/HRT risks and benefits, you must chart a patient and careful course through this passage of your life.

Two valuable telephone "hot lines" have been set up as cancer information services. They are staffed by cancer information specialists (not physicians) who can answer more of your questions about cancer or will be happy to send you free information about this important topic. The toll-free hot line provided by the National Cancer Institute is 1-800-4-CANCER; the number for the American Cancer Society is 1-800-ACS-2345. Both lines are staffed by pleasant, knowledgeable people, who are very helpful. For example, you can request printed directions on how to do breast self-examination or learn what the American Cancer Society's recommendations are for how often we should have mammograms.

For now, mammography remains the best way of detecting signs of breast cancer.

Susan M. Love, M.D., with Karen Lindsey
Dr. Susan Love's Breast Book
Addison-Wesley, 1991

CHAPTER 8

Does Getting Older Mean Getting Shorter? What You Can Do About Osteoporosis

Last summer while I was on vacation, I bumped into a woman I hadn't seen for a few years. When we had first met in a water aerobics class, she was straight and tall. When we met again, she appeared heavier, shorter, and her face seemed to emerge from her clavicle. I couldn't see the stem of her neck anymore. Always cheerful, she greeted me with, "I bet you didn't recognize me. I've gained a few pounds!"

I, however, was not looking at her pleasantly plump frame or at her still-pretty face. I was trying not to stare at the dowager's hump that was curving her back and making her neck seem to disappear. I could not believe my eyes.

We spoke for a while and then she said, straight out, "Did you notice—I'm three inches shorter now and plenty scared about it? But what can I do? It's my family's legacy to me!"

Over the next few days, I thought about this encounter. I was fairly certain that this woman had osteoporosis. I wondered if she knew. The results of a 1991 National Family Opinion Survey conducted among a thousand women, ages forty to sixty, indicated that nearly all of the women surveyed (95 percent) had heard of osteoporosis and that most of them were somewhat familiar with its effects.

It was no surprise that the postmenopausal women in the survey knew more about osteoporosis than the premenopausal women or that approximately one third of the women who had heard of osteoporosis had discussed it with a physician (either their general practitioner or their gynecologist). Their physicians usually recommended one or more of the following: taking calcium supplements (39 percent); taking estrogen (36 percent); exercising (28 percent); eating a healthier diet (19 percent); and eating more calcium-rich foods (11 percent). It is interesting to note that the women who had had these discussions with their doctors were more often those with a family history of the disease, or who already knew that their early menopause could increase their risk of osteoporosis. These were often "take charge" kinds of women who were already taking some steps to protect themselves from their potential risk, such as improving their diets, taking additional calcium, increasing their exercise programs, and/or taking ERT.

When the 43 percent of the women surveyed who had taken ERT at some time were further studied, it was found that these women usually had become familiar with osteoporosis through their physicians; they were more aware of some of the long-term benefits and of the risks of ERT; and they indicated that they believed that the overall benefits outweighed the risks.

Most of the women in the survey understood that osteoporosis causes bone degeneration or shrinkage, and more than 90 percent knew that calcium intake, proper diet, and exercise were linked to the prevention of osteoporosis. More than half of the women also considered estrogen to be an effective preventive measure against osteoporosis. Most important, these women did not accept osteoporosis as an inevitable by-product of aging and knew that they had to take action in order to prevent osteoporosis. Although some women were unclear about the type of testing used to detect bone loss, 83 percent felt that testing for osteoporosis in women over forty should be covered by medical insurance (as of this writing, it often is not covered).

Take a few moments to consider where your knowledge level about osteoporosis might fit into these study results. Now, let's look at the facts.

The startling facts about postmenopausal osteoporosis are that it currently affects more than twenty-five million Americans (mostly women); that one out of two women over the age of fifty may be at risk; and that it costs up to ten billion health care dollars each year. Those are the plain facts; here are the *pain* facts.

Osteoporosis is a debilitating bone disorder. Its first symptoms usually are a reduction in a woman's height and the start of a dowager's hump: that stoop or curvature of the spine that is commonly caused by fractured vertebrae resulting from thinning bone. A sharp snapping sound or sensation when lifting or bending, or a sudden pain in the lower back often heralds the diagnosis of osteoporosis. It is fractures of the hip that are the most serious, however; 20 percent of such fractures may even prove fatal due to the medical complications, such as blood clots or pneumonia that may follow the fracture. This figure is what makes hip fracture the twelfth-leading cause of death in postmenopausal women. Equally devastating is the fact that two thirds of the women with hip fractures never fully recover or can resume their previous life-styles and often suffer greatly because they can no longer take care of themselves.

Osteoporosis has been appropriately named the silent thief, because it creeps stealthily upon us, quietly scavenging our bones, with visible symptoms appearing only after the disease has reached an advanced stage. The good news is that we can change these statistics, if we choose to make the life-style changes that can protect us, and if we avail ourselves of the preventive medical care that can help us.

According to Dr. Charles H. Chesnut III, director of the Osteoporosis Research Center at the University of Washington Medical Center in Seattle, "The only proven preventive for postmenopausal osteoporosis is estrogen replacement therapy. Many physicians will recommend a combination of therapies, including exercise and a diet rich in calcium in conjunction with estrogen replacement."

Scientists used to believe that bone lost was bone gone forever, but studies conducted in the last few years among older retired adults now indicate that with exercise, new bone may be laid down even in individuals in their eighties. So, the great news is that it's never too late to try and build bone or to conserve bone and prevent further loss.

Questions and answers concerning the healthy lifestyle habits that may prevent osteoporosis or slow its progress are in Chapter 13. This chapter will cover the important basic facts about osteoporosis, its diagnosis, and its treatment.

83. WHAT IS OSTEOPOROSIS?

Osteoporosis literally means porous bone. It is the most common and potentially the most debilitating bone disorder in the world. This disorder is characterized by loss of bone mass, which causes us to have bones of reduced density and strength. These thinner, more brittle bones greatly increase our risk of fractures. To some degree, the nature of bone itself makes it vulnerable to this disease. Bone is growing, living tissue. The loss of bone mass can occur during the body's natural lifelong process of bone remodeling. It works this way: There are voracious Pac-Man–like critters in your body called *osteoclasts*. These are programmed to "gobble up" and get rid of old bone tissue in a process called resorption. There are also worker-critters, called *osteoblasts*, whose job it is to busily build new bone tissue to replace the old. In nature's tug-of-war, it is important that the work of the gobblers and of the builders stays in balance, so that bone is remodeled (or renewed), but is not lost. Unfortunately, as we age this is not always possible. Usually up to age eighteen, the builder-workers are ahead and our bones grow. Then, at around age thirty-five, the balance changes and most people begin to lose bone, at the rate of about 1 percent per year. Around menopause, when estrogen levels begin to drop, women lose bone at an accelerated rate, as much as 3 percent of their total bone density in each of the first five years after menopause (and about 1 percent per year after that).

84. What are the risk factors associated with postmenopausal osteoporosis?

The National Osteoporosis Foundation has identified the following risk factors that may contribute to the development of osteoporosis: a family history of osteoporosis; an early menopause (either natural, usually meaning before age forty-five, or surgical, due to removal of the ovaries); being Caucasian or Asian; a sedentary life-style; and a chronically low dietary calcium intake. Other risk factors that have been implicated are thinness in women, heavy use of alcohol or caffeine, and smoking.

85. WHAT CAN I DO TO PREVENT OSTEOPOROSIS?

Learn your family history. If this cannot be supplied by members of your immediate family, try to find someone else, such as an aunt who might fill you in on how the women in your family age. Start adopting healthy life-style habits *now*. If we begin early in our lives, we can greatly minimize our risks. Make sure you're doing your share of weight-bearing exercises, that you're getting enough calcium, and that you limit your alcohol consumption, because alcohol can lessen your bones' ability to hold onto their calcium. Quit smoking! Smoking can bring on an earlier menopause, reducing your estrogen production sooner. Discuss osteoporosis prevention with your physician. Learn whether you are a candidate for ERT. In October 1991, the Food and Drug Administration approved the transdermal skin patch for use in the prevention of postmenopausal osteoporosis. The approval of the patch was based on a two-year research study conducted at the Mayo Clinic in Rochester, Minnesota, which demonstrated the patch's long-term benefits in preventing postmenopausal bone loss that can lead to fractures. An estrogen pill, Premarin, is also approved for the prevention of postmenopausal osteoporosis.

86. HOW IS POSTMENOPAUSAL OSTEOPOROSIS USUALLY DIAGNOSED?

Your physician will probably want to know your complete medical history and your family's medical history. That information, plus a thorough physical examination and radiological testing of your bone density, if indicated, should provide the foundation of a good diagnosis. However, if fractures have already occurred because of brittle porous bones, the disorder is already under way and the diagnosis is pretty clear. See your doctor!

87. WHEN SHOULD A WOMAN START TAKING ESTROGEN IN ORDER TO PREVENT POSTMENOPAUSAL OSTEOPOROSIS?

That is a question that needs to be answered by your own physician. It is known that women undergo an accelerated rate of bone loss at menopause, when the estrogen production "factory" in our ovaries has slowed down considerably or has gone out of business altogether. But the risk for osteoporosis may begin even earlier. We usually reach our peak bone mass at around the age of thirty-five, and so our risk may begin to accelerate even then. However, it is at menopause that our bone-builder critters lose ground, because they thrive on estrogen. In fact, after the decline in estrogen production that occurs at menopause (or after surgical removal of the ovaries), women generally experience between three and seven years of very rapid bone loss. According to Dr. Chesnut, "Bone loss may begin in adults when they are thirty-five to forty years old at the rate of about one percent per year. Once a woman reaches menopause, her bone loss accelerates two to three percent annually on top of that." Then it slows. ERT can compensate for our natural decline in estrogen production, preventing the accelerated bone loss that otherwise usually begins at menopause. Obviously, ERT is most effective when taken early in menopause, before much bone thinning has occurred, but taking it later can be helpful as well.

88. WHAT KINDS OF TESTS ARE USED TO DETERMINE IF I'M LOSING BONE?

There are several kinds of bone measurement tests in use. They are reliable and painless and usually can be performed in fewer than thirty minutes. The most common tests used today are dual photon absorptiometry and dual X-ray absorptiometry. Purported to be the most definitive sophisticated bone mass tests, these are used predominantly to test bone in the hip and spine. The CAT scan (computerized axial tomography), and single photon densitometry, most commonly used to evaluate the bones in the wrist, are two other radiologic tests that are used to test bone. There are also laboratory tests that check the amount of calcium in the urine. If you need to be tested, the test that your physician chooses for you would, of course, depend on your level of potential risk.

It is difficult to give the costs of these tests because

they vary greatly throughout the United States. For example, the dual photon test varies from fifty dollars to two hundred fifty. In addition, the equipment for these X-ray tests is expensive and, thus, not easily available and not all insurance plans cover them.

89. WHAT IS THE BEST TREATMENT FOR POSTMENOPAUSAL OSTEOPOROSIS?

There are a number of treatments that are presently undergoing clinical testing (studies with patients). These studies may tell us whether ERT started after bone loss can actually help renew bone rather than simply slow further loss. The only products approved by the FDA to be marketed for the treatment of postmenopausal osteoporosis are estrogen and salmon calcitonin injections. In our own body's chemistry, calcitonin is the chemical that enables us to absorb calcium. If we are lacking sufficient calcitonin, it may be because of our low level of estrogen. (Calcitrol, a type of vitamin D currently being studied in New Zealand, may be an alternative treatment in the future.)

90. DOES BODY WEIGHT PROTECT AGAINST POSTMENOPAUSAL OSTEOPOROSIS?

It is true that thin, delicate, small-boned women have greater potential risk for osteoporosis than heavier women, but there are no guarantees that weight will protect us, and there are a number of other risk factors that must also be considered, such as smoking or a sedentary life-style.

91. HOW MUCH CALCIUM SHOULD A WOMAN TAKE DURING PREMENOPAUSE?

As of this writing, the recommended Daily Allowance (RDA) for calcium for adults is 800 milligrams, yet studies show that most women consume around 500 milligrams, which is too little. Moreover, there are individual differences in our ability to absorb calcium. New medical thinking recommends that premenopausal women get about 1,000 milligrams of calcium each day and increase that to 1,500 milligrams after menopause, if they are not on ERT. Eating calcium-rich foods such as dairy foods, canned salmon (with the bones), sardines, mackerel, tofu, nuts, calcium-fortified orange juice, or certain leafy green vegetables that are high in calcium (kale, broccoli, turnip, collard, mustard, or dandelion greens) can boost your intake. Be sure to check with your physician if you are considering taking calcium supplments.

92. CAN I REALLY REPLACE CALCIUM WITH ANTACID TABLETS? IF SO, HOW MANY SHOULD I TAKE PER DAY?

Certain antacid tablets contain a form of calcium carbonate that may supplement calcium intake. I take one that provides 200 milligrams of calcium per tablet. The dosage that you need depends upon how much calcium you are getting from your diet. I encourage you to work this out with your doctor and check the antacid tablets out with your pharmacist.

93. IS THERE A SPECIAL TIME OF DAY TO TAKE CALCIUM SUPPLEMENTS?

I always take my two antacid pills just before bed. In that way I have spaced my calcium consumption over the course of the day, having had some calcium-rich foods at each meal. However, I recently learned that it is better to take calcium supplements with food, in order to improve absorption. There does not appear to be a hard-and-fast rule or preference at this time.

94. HOW EFFECTIVE ARE CALCIUM PILLS AFTER THE AGE OF SIXTY?

There is no cessation of the need for calcium intake after the age of sixty. Currently, the RDA *does* drop to 1,000 milligrams for women over sixty who are on estrogen therapy, because estrogen has proved to be effective in maintaining bone. Nonetheless, the need for calcium-rich foods in our diets—and perhaps for calcium supplementation—continues in later life.

In support of our bones, the National Osteoporosis Foundation (NOF) was founded in January 1986. It is a nonprofit health organization dedicated to reducing the prevalence of osteoporosis. The NOF is the foremost resource for women and for health-care professionals and organizations that are seeking the most recent medically sound information concerning the causes, the prevention, and the treatment of this disease which robs us of our strength and of our lives. Information on how to contact the NOF is listed in the Appendix.

Just as a building depends on its solid construction and on the strength of its beams and its girders to keep it upright and strong, so we depend on the strength of our bones to maintain our skeletons and, ultimately, our good health throughout life.

We shape our buildings: Thereafter they shape us.

Sir Winston Churchill
Time, September 12, 1960

CHAPTER 9

What Is Estrogen's
Effect upon Heart
Disease?

Heart attacks were traditionally believed to be the illnesses of men, caused, some thought, by the stresses of
their work lives. That might have made some sense.
Heart attacks afflicted many men in their fifties and
younger, while women lived longer and were rarely
afflicted. Then, as men and women gradually gained
longer lifespans, (women's still exceeding men's), women began to experience heart attacks in greater numbers, too.

Some said it was because we women had ventured
out into man's world, clambering successfully over some
of the barriers that had sealed off the world of work, and
so we had begun to suffer the same kinds of stresses that
were causing coronary heart disease (CHD) in men. That

always seemed to me to be a fallacious theory, inasmuch as I believe that there is no work more stressful than the combined roles of wife, mother, daughter, homemaker, and those dozens of other roles that millions of women traditionally fulfilled prior to entering the workforce in droves.

Obviously, I wasn't the only person who felt that way. Researchers began to apply themselves to investigating why some women began to experience heart disease around their mid-sixties or early seventies, approximately ten years later than men. It was thought that menopause and our lack of estrogen production might be the culprit. So after many decades of research into CHD with clinical studies done only with men, research into CHD in women began.

Women rarely experience heart attacks prior to menopause. Research today suggests that estrogen offers some protection against the blockages that can develop in the blood vessels and cause heart attacks or strokes. Studies seem to point toward estrogen's action upon the serum lipids in the blood. Early data demonstrates that estrogen increases the "good" cholesterol or high density lipoprotein (HDL) and decreases the "bad" cholesterol or low density lipoprotein (LDL) in the blood. This affords the blood smoother passage through our blood vessels, clearing out the plaque build-up that can block its way.

A heart attack occurs when the blood is blocked from reaching the heart; a stroke occurs when blood cannot reach the brain. Heart attacks and strokes take the lives of more than one million women in the United States and Europe each year. Heart disease is the number one killer of American women, with heart attacks taking the lives of two hundred fifty thousand women and another ninety thousand claimed by stroke. Research to learn how to reduce these numbers is vital.

There are a number of research programs in progress that examine heart disease in women. One important research program was conducted at Harvard. Thirty-two thousand nurses were studied for four years and led investigators to learn that women who took estrogen appeared to have half of the risk of fatal and nonfatal heart attacks. Many other important studies support this link between estrogen and coronary heart disease, yet this relationship has not been conclusively proven.

In June 1991, the United States Food and Drug Administration (FDA) Advisory Committee on Maternal Health gave ERT its official recommendation. It advised that ERT be made available to "virtually all" postmenopausal women. This committee further ruled that the benefits of hormone replacement therapy are sufficiently clear so that pharmaceutical companies can submit applications for approval of new drugs which provide both estrogen and progestin in a single application.

All of this information is offered up in a singularly confusing scientific environment, one that is confusing to physicians as well as to women. Critics of hormone replacement therapy abound, basing their concerns largely on the fact that long-term effects of the combined therapy are still unknown. That is because most of the research has been done with estrogen alone, although estrogen and progestin therapy combined has been prescribed widely in the United States for the woman whose uterus is intact.

During the summer of 1991, interim results of the Harvard Nurses Study were released suggesting the cardio-protective effect of estrogen. Now at the ten-year halfway point in this ongoing study of more than a hundred twenty thousand nurses questioned by mail at regular intervals, the results so far have shown that women who go through early menopause (before age forty) or through surgical menopause (before forty) have a significantly higher risk of heart disease. This finding appears to link the functioning ovary and its hormones to heart health.

Of course, there are also many studies on the heart-healthy effects of good nutrition, exercise, stress reduction, and quitting smoking. These life-style factors are discussed in Chapter 13, and they are life-style choices that you can and should make now.

The question of the effect of ERT on coronary heart disease is not fully or finally answered—although the implications are becoming clearer—so this chapter can address the women's questions only by sharing the information that appears to be known at this time.

95. DOES ERT REDUCE THE RISK OF HEART ATTACK?

Even though research findings point to a significant positive relationship between estrogen use and a reduction in heart disease, no estrogen product has been approved at this time by the FDA for the prevention of heart disease in postmenopausal women.

96. ARE THERE ANY LARGE-SCALE STUDIES UNDERWAY CONCERNING CHD AND ERT?
The Nurses Cohort Study of Harvard reached its halfway point in 1986. The results of that study so far indicate a strong association between estrogen deficiency and CHD risk. I would watch for further results from that study. The first major study funded by the government was begun in 1990. It's a three-year, ten-million-dollar study on the effects of hormone replacement therapy on menopausal women, conducted under the auspices of the National Institutes of Health (NIH). You may see or hear it referred to as the PEPI study—the acronym stands for the Postmenopausal Estrogen-Progestin Intervention Trial. Due to be completed in the early 1990s, the results should help scientists better understand the role of hormone therapies in CHD prevention. Dr. Bernadine Healy, the director of the NIH, is requesting funds for other large-scale research projects concerning women's health issues. These sweeping studies are long overdue, and the results will merit our careful attention.

97. WHAT IS THE RELATIONSHIP BETWEEN SMOKING, MENOPAUSE, AND HEART DISEASE?
Smoking may hasten the onset of menopause by five to

ten years. If estrogen deficiency is proven to be a risk factor for CHD (and research at this time points in that direction), then an early menopause may increase your risk of heart disease.

98. I AM FACED WITH A HYSTERECTOMY IN THE VERY NEAR FUTURE. DOES THIS INCREASE MY RISK FOR HEART DISEASE?

Maybe not, if you can keep your ovaries. Prior to menopause, the ovaries continue to produce the female hormones estrogen and progesterone, even though the uterus is gone. It is estrogen deficiency that may be a risk factor for CHD in women. More information about hysterectomy is included in Chapter 4.

99. WHAT CAN I DO TO REDUCE MY RISK FOR CHD?

Stop smoking. Cut down on alcohol. Get a comprehensive physical examination, have your blood pressure, cholesterol, and glucose tolerance checked, reduce the fat in your diet, perform at least twenty to thirty minutes of aerobic exercise at least three times a week, and try to reduce the negative stresses in your life whenever possible. Discuss with your physician whether you should consider ERT.

100. DO BIRTH CONTROL PILLS RAISE OR LOWER MY RISK OF HEART ATTACK?

Birth control pills, or oral contraceptives, contain a higher dose of estrogen than is used in postmenopausal ERT (up to ten times as much). The use of birth control pills is considered a coronary risk factor because of the high dosage. New lower dose "minipills" are available. Ask your doctor about whether you should consider them.

101. WILL ERT PROTECT ME FROM HAVING A HEART ATTACK?

According to an article in the prestigious *New England Journal of Medicine* on October 24, 1989, coauthored by leading investigators at Brigham and Women's Hospital and Harvard Medical School: ". . . postmenopausal women who take estrogen generally have lower rates of cardiovascular disease than women of a similar age who do not, possibly because estrogen has favorable effects on plasma lipoprotein levels, which are risk factors for cardiovascular disease." The article indicates that if the findings are borne out, "the risk of cardiovascular disease may decrease by more than 40 percent" in women taking ERT, and it concludes, like so many other articles, by calling for a clinical trial to demonstrate definitively the effect of estrogen upon cardiovascular disease. Even though research findings point to a significant relationship between estrogen and heart disease, remember that there is no estrogen product approved at this writing by the FDA for the prevention of coronary heart disease in postmenopausal women.

102. I'M ON HRT. DOES PROGESTIN CANCEL OUT SOME OF ESTROGEN'S CARDIO-PROTECTIVE EFFECT?

Research indicates that progestin may counteract some of estrogen's suggested positive effects on CHD. Yes, it's confusing. It's so confusing, in fact, that at a meeting of the FDA Advisory Committee on Fertility and Maternal Health Drugs in June 1991, it became crystal clear that there are actually no progestins approved by the FDA for use in HRT, not even those that are currently in use in HRT. Again, the cry for large-scale studies to answer these questions definitively is heard loud and clear.

103. SO WHAT DO I DO WHILE WAITING FOR THE STUDIES TO TELL ME WHETHER I SHOULD CONSIDER ERT TO PROTECT ME FROM HEART DISEASE?

You're in the company of millions of women who have to make a decision about hormone therapy. You need to study the facts and then come to some kind of interim conclusion with your physician. On the plus side is the fact that women don't usually suffer CHD or heart attacks before menopause. It is in the decade *after* menopause that we catch up to men in this regard. It appears as if prior to menopause our own natural estrogen may protect our heart by raising our level of HDL cholesterol, which is thought to sweep through our arteries and cleanse them of plaque or even prevent these fatty blockages from forming on the walls of our arteries. Other studies, including the ongoing, long-term Harvard study mentioned in questions 95 and 96, suggest that taking ERT postmenopausally may cut the risk of these women dying of a heart attack in half. By 1986, some forty-eight thousand women in the study of one hundred twenty thousand nurses had become postmenopausal. The interim report issued on the study noted that the incidence of heart attack among the group on estrogen was significantly lower than that of the group who had not taken the hormone. Yet, many physicians still agree that the use of ERT as preventive medicine for postmenopausal women is still a matter of debate.

I personally recoil at having to include such ambiguous information in this book, but on the subject of estrogen and heart disease, there are no concrete answers—yet! Some very encouraging reports are emerging, however. We are getting closer every day to the answers. In the meantime, you and your doctor need to explore what is right for you in this regard and stay alert for future research findings as science attempts once and for all to get to the heart of the matter!

CHAPTER 10

What Are the Alternatives to Estrogen Replacement Therapy?

A woman from Sweden was sitting on the beach in Key Biscayne, Florida (where I live and write in the winter). Upon learning from a mutual friend that I was working on a book about menopause, she came over to my "spot" to talk to me. I'll call her Ursula. Ursula wanted to discuss all the natural ways to stay healthy and relatively symptom-free. Ursula had a "mindset need," not a medical need, to pass through menopause without medicine. She is a health-food devotee and a daily exerciser who includes strength, flexibility, and aerobic regimens in her fitness program and who takes a day off from exercise reluctantly, if she is obliged to do so.

Ursula was interested only in homeopathic medicine. She related what she had been doing to follow her own carefully designed route through her symptoms and to surmount each obstacle to feeling good as it crossed her path. Ursula looked great! She said she felt great, too. Yet, in light of all the talk of ERT and the many different approaches her friends had taken to ward off menopausal symptoms, she expressed some ambivalence about her chosen methods.

We talked about the fact that not all women *need* to take estrogen; that some women *cannot* take estrogen because of current or chronic medical conditions; and that many women, like Ursula, *choose not* to take estrogen because they do not want to "interfere with nature." Still others might like to take ERT, but are fearful of doing so. With those categories in mind, let's look at some alternatives to ERT.

Natural alternatives to ERT abound. According to Dr. Susan Lark's *The Menopause Self Help Book*, vitamins and minerals obtained through appropriate and careful dietary selections—sometimes with supplements added—can help to alleviate and may even prevent many mild menopausal symptoms. For example, she writes, potassium and magnesium aspartate can improve energy levels in menopausal and postmenopausal women. That is very significant information, considering that 56 percent of the women who completed our program questionnaires indicate that they have experienced lower energy levels since the onset of menopause.

A word of caution here, however. If you wish to explore how vitamins and minerals may relieve some of your symptoms, be sure that you take the time to understand their properties and what they can do for you, and discuss your ideas with your physician or with a qualified nutritionist. Appropriate dosage is essential as well, since overdosing can have toxic effects. This is a perfect example of a case in which too much can be as bad, or even worse, than too little. Remember, too, that these recommendations may not protect you from postmenopausal osteoporosis. Your physician can direct you in this regard.

I know that Vitamin E helped me to control hot flashes even though there are no conclusive scientific studies to support that fact. I took 800 milligrams per day—400 in the morning and 400 before bed—for a number of years to stop the night sweats that I thought were stress-induced (that's years before the realization of menopause crashed upon me in 1985). A friendly health-food-store owner had suggested Vitamin E to me. She said I could consider taking up to 1,200 IU (International Units) per day, but no higher. I checked with my doctor, who said, "Why not try it?" I still take Vitamin E each day, limiting it to 400 IUs in the morning with breakfast—I no longer need more than that.

Dr. Lark credits Vitamin E with the ability not only to relieve hot flashes and vaginal dryness, but also to alleviate some psychological symptoms, such as mood swings, fatigue, and anxiety, when taken along with other appropriate nutrients like potassium, magnesium, the B vitamins, and bioflavonoids. Vitamin E has been studied for its ability to reduce breast cysts. Some studies, such as the one at Johns Hopkins School of Medicine reported in the *Journal of the American Medical Association* way back in September 1980, suggest that it is helpful in this regard. Other studies suggest that Vitamin E may be useful in improving certain skin conditions, osteoarthritis, and heart disease. The best sources of Vitamin E include wheat germ, lettuce, and green peas. Other good food sources include the following: other green vegetables (asparagus, cucumber, kale); wheat germ oil (and oils made from corn, safflower, sesame, soybean, and peanut); fish (haddock, mackerel, and herring); meat (lamb and liver—but be wary of the high cholesterol content of liver); grains (brown rice and millet); and mango, a tropical fruit. However, megadoses of anything can be risky, so caution is needed. Vitamin E usually is not recommended for anyone with high blood pressure, diabetes, or a rheumatic heart condition, and it should not be taken at all by anyone who is taking digitalis. Also, there is some concern on the part of physicians that too much Vitamin E could cause liver problems.

Vitamin B deficiency is not uncommon in women who have taken birth control pills or who are on ERT. The synthetic hormones can deplete the B vitamins. This deficiency may trigger fatigue, depression, emotional instability (mood swings), memory lapses, or loss of libido, because the whole family of B vitamins work together to perform vital metabolic functions in our bodies, which includes stabilizing the chemistry in our brains. Vitamin B-complex is considered to be an antistress compound. The one I've taken for years includes Vitamins B-1 (thiamine); B-2 (riboflavin); B-3 (niacin); B-5 (pantothenic acid); and B-6 (pyridoxine), together with Vitamin C. Some women have told me that B-complex gives them some relief from migraine headaches. Vitamin B-6, in mild doses, such as 50 milligrams, usually works as a natural diuretic, counteracting water retention that makes so many of us feel bloated or uncomfortable. Other natural diuretics include cranberry juice, kelp, watercress, and parsley. Reducing salt and spicy foods helps too. The B vitamins can also be supplemented through your diet with greater consumption of beans, whole grains, and liver (again, liver's high cholesterol content makes it just a once-in-a-while food).

Vitamin C can help as well. It is considered an antistress vitamin possessing calming effects. It is also a deterrent to excessive menstrual bleeding, a healer of wounds and burns, and a maintainer of collagen, which is the main supportive protein of your skin, tendon, bone, cartilage, and connective tissue. Vitamin C is found abundantly in fruits and vegetables. Some of the best vegetable sources are broccoli, brussels sprouts, cabbage, cauliflower, and most greens. The best fruit sources of C are cantaloupe, grapefruit, oranges, strawberries, mango, and papaya. If taken as a supplement, time-release Vitamin C is best.

Calcium, in addition to helping to maintain healthy bone, is also recommended for coping with emotional stress. Vitamin D assists in the body's absorption of calcium, and a sufficient amount of Vitamin D is usually acquired through our exposure to sunlight. The best food source of Vitamin D is salmon, followed by mackerel, sardines, and tuna. More detailed information about bones, osteoporosis, and calcium supplementation is provided in Chapter 8.

There are a number of herbs, such as chamomile, blackberry root, passionflower, and evening primrose oil, which when taken in moderation may alleviate some menopausal symptoms. This makes sense, because some plants contain estrogen. Caution is important, however. First of all, herbs can be toxic if ingested in the wrong quantities. It is also possible to have an allergic reaction to them. Ginseng, for example, is a root that is a source of plant estrogen, and it can enhance energy and may relieve hot flashes. Yet it is not often recommended by physicians, because it is like taking ERT without knowing how much you are taking. So if you choose to sip ginseng tea, sip it lightly and slowly and not for long.

Menopausal insomnia is a serious problem for many women with whom I have spoken. Some of the women have offered suggestions of what has worked for them. Many suggest herbal teas—one company even calls its chamomile tea Sweet Dreams—and catnip tea, passionflower broth, warm milk, warm baths, and long evening walks.

Vaginal dryness can be helped with the new water soluble vaginal moisturizers, now available in over-the-counter products, such as Replens and Astroglide. In addition, it should come as no surprise that regular sexual activity can help keep your reproductive organs in good shape (this is covered more fully in Chapter 12).

Kegel exercises can help with bladder control if stress incontinence is a problem. They also can tighten and tone the vaginal area, which may enhance sexual pleasure for you and your partner. Kegel exercises, named for Dr. Arnold Kegel, a surgeon at UCLA who developed them in the 1950s, may be done in one of two ways. Method one is to contract your vaginal muscles as if you were trying to stop yourself from urinating. Hold for a count of five, relax for another count of five, and repeat this sequence twenty times. Method two is to contract and release in quick succession. I not only alternate the releases, I alternate the methods. It's like cross-training, and it keeps me from getting bored. I still do these as many times a day as I can remember to do them. Memory triggers that I use include doing "Kegels" when I stop for red traffic lights, "Kegels" while taking off my makeup at night, and "Kegels" when I put on makeup in the morning. I try to do at least ten sets of twenty each day. Some days, I'm sure I miss. I know that the Kegel exercises have helped me with bladder control, and I believe that they may help keep my pelvic organs functioning well my whole life. I think they're well worth the small effort.

A number of other techniques have provided relief from distressing menopause symptoms for many women with whom I've talked. These include relaxation therapy, visualization, yoga, and biofeedback. Acupuncture and massage are also considered for the relief of some menopausal symptoms. I will briefly review them here, but since these are such fascinating and unique therapies, you may wish to learn more about them through the many books that are available on these subjects. A partial listing is included in the Appendix.

Relaxation therapy techniques can help to relieve stress and tension and, in that way, combat some menopausal symptoms. Relaxation therapy is a process through which you systematically relax your body and your mind from head to toe. Visualization is another relaxation technique, but one in which you create your own desired atmosphere and environment. For example, in order to counteract hot flashes, you might envision yourself ice fishing on one of the magnificent lakes around Minneapolis. You pull up the hood of your down parka and tie your wool scarf a little tighter around your throat. The wind is cold, yet it is sunny and you are at peace waiting for a fish to nibble at your line, which has been dropped down the hole in the ice in the corner of your ice house. For many women, visualization works wonders!

Yoga combines deep breathing, meditation, and physical exercise. It is believed to focus one's attention and calm the mind. When combined with good nutrition and healthy life-styles, it can be helpful in alleviating the problems of menopause and aging. Yoga, when practiced correctly, also can enhance your muscle and joint flexibility greatly and help to maintain your skeletal health.

Biofeedback is a painless relaxation technique that involves being hooked up to a machine that can help to train the mind to grab control of the body's mechanisms (such as heart rate, muscle tension, or skin temperature) which are usually the province of the body's automatic response systems. The feedback comes from meter readouts or tones that tell you whether there is an increase or decrease in what you are attempting to control. Women have told me that they have been able to stop their hot flashes through biofeedback.

Acupuncture is an ancient Far Eastern healing art that involves the insertion of needles at certain points on the body's meridians for the purpose of unblocking our energies. In my travels, I have encountered a few women who have had good results using it to relieve their hot flashes. In the United States, acupuncture may be lawfully practiced only by licensed physicians using disposable needles. In this, as in all other procedures and practices, check the practitioner's credentials carefully.

Massage, involving acupressure and trigger-point therapy, is something I myself use on a regular basis. Therapeutic massage brings me relief from all kinds of symptoms, from muscle strains due to sports or exercise, to headaches and stiff neck (my personal stress spot), as well as removing toxins from my body. Massage such as

I undergo can be done only by a licensed massotherapist, in most states. It involves the placement of digital pressure at the same points on the body's meridians as is done with the insertion of needles in acupuncture, for the same purpose of releasing the body's energies. Trigger-point therapy pinpoints congestion in the muscles and works with digital pressure to relieve it. There are a number of other forms of massage therapy that work well for other women. Some women enjoy Swedish massage, Shiatsu massage, or foot reflexology, just to name a few. If massage therapy interests you, I suggest that you take the time to learn about the various forms of massage and what they may do for you and to find a qualified licensed therapist with whom you are comfortable. If you try one kind of massage and it is not to your liking, experiment and try another.

Just as there are a number of simpler means of trying to improve sleep, such as drinking warm milk, so there are some equally simple things to do to avoid or reduce the effects of hot flashes. Women report success with iced drinks, cold showers, dressing in layers and peeling clothing off as necessary, avoiding alcohol, caffeine, and spicy food, and limiting unnecessary stress whenever possible. A friend of mine can literally bring on a series of hot flashes just by becoming nervous and upset. The reverse is true for many of us who pursue activities for their calming effect.

There are also some medical alternatives to ERT. Tranquilizers, such as Valium, may be prescribed to reduce the anxiety and stress that can be associated with menopausal symptoms. They work to diminish some symptoms, such as hot flashes or insomnia, because they suppress the function of the hypothalamus, but they can become habit-forming with extended use. In addition, your body may build up a tolerance to them, which would then necessitate increased doses over time. Sedatives also decrease irritability in the autonomic nervous system, but these should be considered only for temporary help because they are addictive, and they should be used with caution.

If you can't take estrogen, some physicians will prescribe progestin alone, a strategy that may offer some help for your symptoms. Clonidine is another drug that may be tried. It is an antihypertension drug that may provide hot flash relief. Some physicians prescribe Bellergal tablets, which contain phenobarbital. Bellergal is most often used as an antispasmodic drug. It, too, may reduce hot flashes. Remember: All drugs can have side effects that need to be considered and understood.

104. HOW MUCH CALCIUM SHOULD A WOMAN TAKE TO PREVENT OSTEOPOROSIS?

As of this writing, the Recommended Daily Allowance (RDA) for calcium is 800 milligrams. If you are premenopausal, you should take in at least 1,000 milligrams per day. Four 8-ounce glasses of milk would provide 1,200. (Skim milk is as calcium-rich as whole milk and is virtually fat-free.) After menopause, because your body has become less efficient in its ability to metabolize calcium, your requirement increases to 1,500 milligrams a day (unless you are on ERT, in which case 1,000 remains satisfactory). Calcium supplements are recommended for women who do not or cannot (because of lactose intolerance) consume enough calcium in their diet. The best food sources of calcium are milk (skim or whole), yogurt, cheese, raisins, oysters, salmon (with the bones), sardines, mackerel, tofu, calcium-enriched orange juice, and many of the leafy green vegetables. Remember, while estrogen is the only proven preventive for postmenopausal osteoporosis for women at risk, many physicians will recommend a diet that is rich in calcium in conjunction with ERT.

105. VITAMIN E SEEMS TO HELP CONTROL MY HOT FLASHES. IS THERE ANY SCIENTIFIC PROOF THAT IT WORKS? A number of studies seem to indicate that Vitamin E can help alleviate hot flashes and some of the psychological symptoms of menopause, as well as relieve some of the vaginal dryness and soreness that women may experience. However, none of these studies has proved conclusively that Vitamin E works or exactly how it works.

106. IS DRY AND ITCHING SKIN A PART OF MENOPAUSE? It is a part of menopause and a part of aging, since as we age our skin becomes thinner and dryer. There are a number of ways to relieve these symptoms and to improve the look and feel of your skin. Apply a skin mois-

turizer immediately after your bath or shower to trap moisture in your skin. Take tepid baths or showers instead of hot ones, which are more drying. Add bath oil to your bath water and luxuriate in a long soak.

107. WHAT CAN I DO TO REPAIR MY BLOTCHED SKIN, OR AT LEAST TO MAKE IT LOOK BETTER?

Our skin changes somewhat as we age and these changes are more acute if over the years we have exposed ourselves to damage from the sun. So, in addition to getting dryer or itchy, our skin may wrinkle as we age and may also change in texture and tone. In addition to those brown age spots, the skin on our faces may become blotchy and uneven as a result of broken capillaries and hormonal changes. These changes may occur on our hands, arms, and elsewhere on our bodies as well. There are a number of procedures that a dermatologist can perform to rid you of some of the spots. There is success with freezing off the brown spots, a process that uses electrical current to seal off the broken capillaries, and with applying Retin A cream or gel to rid us of fine-line wrinkling. Today, new treatments have been developed using lasers to repair some of these skin problems. If these medical procedures are not an option for you, you can also conceal changes in your skin with appropriately chosen makeup products.

108. WHY DOES MY SKIN SUDDENLY SEEM TO BE A DIFFERENT COLOR AND WHY DOES MY MAKEUP LOOK HARSH?

Our skin lightens as we age, so it is important that you change your foundation makeup and blusher to complement the change. Above all, it's never too late to stay out of the sun or to go into it fully protected with sun block and clothing. Many cosmetic companies are now formulating makeup products that include sunscreen as one of the active ingredients. Vitamins A and E are helpful in the support of the skin, and so is getting enough of the essential fatty acids in your diet. The so-called omega-three fatty acids are primarily found in fish and are believed to be heart-helping through lowering cholesterol and triglycerides and by improving blood circulation, which, in turn, helps our skin to look and feel better.

109. WHAT CAN I DO TO HELP THIS FEELING OF BEING BLOATED ALL THE TIME?
Stop shaking the salt shaker. All the salt we need in our diets can be found in fresh vegetables, fruits, grains, and meats. Yet there is so much salt in prepared foods, such as catsup and salad dressings, that we should seek to eliminate salt from our diets whenever possible. Salt can increase bloating from fluid retention as well as increase blood pressure. Read labels and look for products that contain little or no salt. Use herbs and spices to make your food more flavorful. Salt is an acquired taste, so once you cut down, you will notice that salty food is much less appealing to you.

110. WHY HAVE I BEEN TOLD TO CUT DOWN ON MY CONSUMPTION OF SUGAR AFTER MENOPAUSE? I DON'T HAVE DIABETES.

After menopause, we begin to see the negative effects of sugar upon our systems in several important ways. Excess sugar steals our B-complex vitamins and minerals. You will recall that earlier in this chapter I described these as offering calming effects. With the depletion of B-complex vitamins and minerals, we may become more nervous or anxious—a real concern at menopause, when stress may worsen hot flashes and other symptoms. Further, sugar promotes tooth decay and gum disease, both of which can accelerate as we age. Even though diabetes is not your concern now, it is known that excess sugar consumption can be a factor in encouraging adult-onset diabetes and other imbalances in blood sugar, and that both of these conditions can worsen after menopause.

111. I KEEP VACILLATING BETWEEN TAKING ERT AND TRYING TO GO THROUGH MENOPAUSE WITHOUT IT. I'M STILL CONFUSED. WHAT ARE THE MAIN REASONS WHY PHYSICIANS PRESCRIBE IT?

Physicians may prescribe ERT for four main reasons: to alleviate the sudden onset of menopausal symptoms if you have undergone an early surgical menopause (see Chapter 4); to help relieve hot flashes and their nighttime twin, night sweats; to relieve vaginal discomfort—dryness, soreness, and the resulting pain during intercourse (see Chapter 5); and to prevent postmenopausal osteoporosis (see Chapter 8).

It is important that we educate ourselves on our own behalf so that we can make informed choices regarding many aspects of our lives. Our continued health and well-being should be right at the top of our priority list. We should look forward to interesting and fulfilling years after menopause. We not only have the right, but the obligation, to manage our own health care as effectively as we know how.

Knowledge is of two kinds. We know a subject ourselves or we know where we can find information upon it.

Samuel Johnson

CHAPTER 11

How Do I Find the Right Doctor?

"How do I find the right doctor?" That is one of the most frequently asked questions after the programs. It's a good question, but one for which we each have to do our own soul-searching before we begin our quest. Often we bring to the doctor's office a list of unrealistic expectations. We want our physicians to fix whatever is wrong, we want them to fix it fast, and some of us feel that unless we leave with a prescription in hand, nothing was done to help us.

Now that we know the expectations, let's examine the realities. First of all, we need to put more time into finding the physician who is right for us and in working with him or her to achieve a sense of trust and easy communication. Getting to that point, however, is not a matter of luck. It involves work on our part.

When I examined the thousands of questionnaires turned in at the programs, I learned that most of the women who attended were the patients of gynecologists (78.2 percent). First runner-up were general family practitioners (14.6 percent), followed by specialists in internal medicine (5.6 percent). The other few percent fell into a category labeled "other." No questionnaire response indicated that any woman did not have a physician, unless, perhaps, that information was thrown into the "other" category.

According to the questionnaires, a few women expected menopause to be a nightmare (2.4 percent), while others felt it to be unpleasant (15.2 percent), or to be "sometimes good and sometimes bad" (20.4 percent). Most women (two thirds), however, indicated that they didn't know what to expect.

The questionnaires indicated other vast gaps in the women's knowledge about menopause—an interesting finding, given the large percentage of women under the care of a gynecologist, who (one might think) would be the logical provider of information. For example, responding to the question "What was most unexpected concerning the symptoms of menopause," 31.8 percent of the women were surprised by the severity of their symptoms; 35.1 percent were not expecting the symptoms to occur so frequently; 21.3 percent were delighted by the mildness of their symptoms; and 15.5 percent did not expect that they would experience any symptoms at all.

From the questionnaires, I found that women had learned about menopause from friends, physicians, and books in about equal percentages and secondarily from magazines, other media, and family. I also wondered whether other women, like me, had subconsciously chosen to ignore what they knew.

Large numbers of women indicate that they attend the educational programs because they still are not sure how to deal with their symptoms and whether to take ERT or HRT. That's not surprising, as there is some confusion on the part of physicians as well, inasmuch as research concerning hormone therapy and women is still in its infancy when compared with research done on other major physiological changes and medical conditions (such as heart disease), which has focused primarily on men. Some say that this is because scientific research has been held in the hands of men throughout history. Perhaps that's one of the reasons. Whatever the case, research into women's menopausal changes began haltingly only about thirty years ago. Now, fortunately for us—and even more fortunately for our daughters—there is a research race to catch us up.

That lack of concrete research results is part of the problem in finding the right physician to care for us at this important rite of passage. If the pathway to appropriate care is not clearly marked, how can we know if the right route has been chosen? The other part of the problem is that we as women still must embrace the challenge of taking control of our own care. Many women indeed are taking steps to learn more about menopause—and those who will probably have the most useful information are the women who are learning what to expect of this interesting rite of passage while they are still many years premenopausal. They are the first generation of women to approach menopause with the information they need to help them fashion for themselves a first-rate second half of life.

In order to achieve that end, you need to work in concert with your own physician. At the end of this chapter there are twelve tips for you to consider each time you go to the doctor. I hope that they help you maximize the quality of your visits. But before we consider the office visit, let's talk about the patient/physician partnership.

I received a telephone call recently from a woman in a small town in western Illinois. "I read your story in *Managing Your Menopause* and I had to call and tell you that you have saved my life!" she told me.

The woman, whom I'll call Laura, was fifty-four years old, a successful artist, and a twin. She was having such constant and debilitating hot flashes and night sweats that she believed that every facet of her life and artistic career were in jeopardy. She had an important exhibit coming up and could not prepare for it.

Laura said that after she read of my eight weeks of menopausal hell, learned that I had gotten help and had gone on to write and lecture about it, she felt reassured. She knew that for her, too, this bad time would pass.

Laura felt that she was having a difficult time obtaining interested medical care. She explained that she had a large estrogen-fed uterine fibroid tumor that she and her physician had agreed to wait for menopause to shrink, because she refused to undergo a hysterectomy to have it removed. In the face of her refusal, the physician's logical thinking was this: The fibroid was nourished by Laura's natural estrogen. Once the ovary ceased its production of estrogen, the fibroid would shrink. That was true, but Laura discovered that she couldn't bear the estrogen-deficiency symptoms.

When she took her symptoms to the doctor, he said if Laura couldn't live with the hot flashes, they could try hormone replacement therapy. "But the fibroid will grow," she protested. "Then we'll do a hysterectomy and get rid of all your problems," he said. Laura felt as if she were back at square one. She still did not want to undergo surgery. She felt out of control.

I asked Laura about her mother's menopause. She said that when she went to her mother for advice and information, her mother, who is in her eighties, told her "we just don't talk about *that*." When Laura explained to her mother that she felt at the end of her rope and out of control, her mother said patiently, "It's because you're too involved in your work and your civic and social activities. You should rest more. You're simply doing too much!" I recognized her mother's comment as fairly typical of that generation, and I laughed inwardly. My own mother couldn't have said it better!

Interestingly, Laura told me that her identical twin sister was having no problem with menopause. We talked about that. Laura is thin, a healthy eater, and a regular exerciser. Lana, her twin, began to gain weight about ten years before menopause, and Laura thinks that Lana must now be at least forty pounds heavier than she is. Laura had read about estrogen being converted and stored in fat cells and had begun to wonder whether that was the cause of the difference in their menopausal experience. Laura also learned from her reading that in selected individual cases, carefully monitored hormone replacement therapy can be given for a short time even when a fibroid is present. That was another piece of information that she felt "saved her life." So Laura is making an appointment with a well-known menopause specialist in the large metropolitan area where her next exhibit will be held, and she's going to see him knowing exactly what she wants to ask him about. That's 90 percent of solving the medical puzzle: knowing what you want to know *before* you go.

Laura promised to let me know what happens. In the meantime, I know she is again taking control of her life. That's what is important. That's why *you* have to find the right doctor for *you*.

112. How can I get my doctor to take me seriously?

First of all, take yourself seriously. Second, make sure your appraisal that he or she is not taking you seriously is accurate. I have no intention of appearing glib in saying this. Not wanting to fill the stereotype of the middle-aged neurotic woman or the dual role of neurotic and hypochondriac, I have been known to recount my symptoms to a doctor more lightheartedly than is appropriate. I remember often working very hard to make sure that I sounded well-adjusted and not too serious about myself. I can certainly understand, in retrospect, why the doctor jollied me along and didn't get too concerned about my problems, either. After all, if I was not exhibiting a serious desire for full answers, why in the world would he offer more information than I appeared to want? So, take yourself and your concerns and questions seriously and present them in a serious and concerned manner.

113. WHAT IS THE BEST WAY TO PREPARE FOR A PRODUCTIVE VISIT WITH MY PHYSICIAN?

Make a list of your questions and use it as the source of your dialogue. If your doctor still seems to be treating you with a "pat on the head, everything will be all right, just trust me, and go on about your life" routine, then I would ask her or him simply and directly whether she or he takes your problems seriously, and, if so, explain that you would like to receive a complete and understandable explanation of whatever it is you want to know. None of this dialogue should be undertaken in anything but a pleasant, noncombative manner. If you still cannot resolve your sense of not being taken seriously, it's time to go doctor-shopping. By the way, if you do make a list for discussion, make it pertinent to that visit, between three and five questions, perhaps. One physician with whom I worked recently suggested that a physician is not prepared with enough time to go through a lifetime of concerns in one visit, unless you have asked in advance for extra consultation time. Developing the right relationship takes time and is created over time.

114. How often should I see my doctor while I am on ERT?

Most physicians indicate that with no other underlying medical conditions, you should be checked twice a year. If you are past fifty (or are younger, but are on ERT) you should have a mammogram and a PAP test once each year and at each visit you should have a pelvic examination, a breast examination, and appropriate blood and urine tests. If your uterus is intact and you experience **any** unusual bleeding, you should call your physician immediately. You should also check with your physician if you experience any other changes in your symptoms or if you have unexpected or unusually uncomfortable side effects as a result of the therapy. Some physicians believe that you should be examined twice each year, whether or not you are on hormone therapy. In addition, do not be afraid to call and ask questions! The same doctor routine applies for women on HRT.

115. What should I do if my doctor doesn't believe in ERT?

Ask your doctor why not. Ask if that is his/her general belief for all patients in the practice, or does it pertain specifically to you. If it is because of your medical condition, make sure you understand and are comfortable with your doctor's reasoning. If this is his or her general opinion and you think you may be helped by hormone therapy, ask your doctor to refer you to someone who may be able to help you, or call your local hospital. Many hospitals today have matching or linking programs which attempt to match a consumer's need with appropriate physician care and will usually give you a list of two to three doctors whom they believe will meet your needs.

116. HOW DO I KNOW IF THE "WRONG" DOCTOR IS TREATING MY MENOPAUSAL SYMPTOMS?
That is a question that comes up at every program. I think that you have to find out if your doctor is interested in treating women before, during, and after menopause. If not, ask for a referral to someone whose primary interest is the care of the menopausal woman. I know that most doctors are referred to us by friends, family, and other physicians, but we do need to stop and think about the referrers and consider whether their medical needs and criteria for care and relationship with a physician are akin to ours, rather than simply accepting a referral unquestioningly. Sometimes we act on others' advice because it's fast and it's easy and we don't have to think too much about it. *Do* think about it! Think about your own comfort level! Think about your desire for a relationship in which you participate; think about what you need to know about your own body, and how it changes.

Pay attention to your symptoms, and know which ones to report to your doctor. If you're not sure, report them anyway.

117. SHOULD A WOMAN CONTINUE TO SEE A GYNECOLOGIST FOR HER YEARLY EXAMINATIONS AFTER HER CHILDBEARING YEARS, OR SHOULD SHE SWITCH TO A GENERAL PRACTITIONER?

There is no rule here. I have a gynecologist whom I see religiously twice each year for the type of examination and tests that I mentioned in the answer to question 114. I consult my internist for other medical issues, conditions, or problems. With the enormous body of new and changing medical information and the many research findings that doctors have to keep up with, I believe in seeing both. However, the field of family practice continues to grow, and physicians in that field usually are trained to handle general medicine, obstetrics, gynecology, and pediatrics, so a family practitioner may fill the entire bill for you. In any instance, if I were happy with my gynecologist throughout my childbearing years, and he or she is interested and competent in the care of menopausal women, I would stay in that practice and be happy that I had found the right doctor for me.

118. WHY DOES MY DOCTOR ALWAYS SAY, "IT'S UP TO YOU," WHEN I ASK, "SHOULD I TAKE ESTROGEN?"

I think you probably have a good doctor who is telling you to educate yourself about your body, about menopause and its symptoms, and about estrogen therapy and its side effects, and to weigh the risk/benefit ratio. Then, you decide! That's really what we women should want, isn't it? More input and decision-making power about what we do with and to our bodies. We can't get that unless we get involved and educated.

119. WHAT'S THE BEST WAY TO NARROW THE LIST OF PROSPECTIVE DOCTORS WHEN I AM SEEKING A NEW PHYSICIAN?

Call your local hospitals and acquire the names of doctors who might meet your needs, and compile a list from coworkers, family, and friends. Then narrow your list by considering those whose names keep coming up, whose hospital affiliations appeal to you, whose training and interests best fulfill your needs, who are in the most convenient location for you, whose office hours and days can work for you, and whose office receptionist is pleasant and helpful when you ask for a consultation/interview/get-acquainted appointment. Find out the charges for these types of office visits and the usual amount of scheduled time allowed for them in advance, so that you are prepared. Then visit the top two or three candidates on your list and make your selection only after you have seen them.

120. HOW WILL I KNOW I HAVE FOUND THE RIGHT DOCTOR FOR ME?

Judge whether you are comfortable in the office with the

doctor and the staff. Were you pleasantly greeted? Was the waiting room comfortable with new and pertinent reading material? Were you made to wait too long? If so, were an explanation and apology offered? Did you feel an easy exchange of information with the receptionist, nurse, and doctor? Was the atmosphere professional/friendly and the environment professional/clean? Were you asked for a comprehensive history? Was the history reviewed? If you arranged to be examined during this get acquainted visit, during the examination, was each step explained clearly to you? Were your questions completely answered? If you received a prescription for medication, were you fully informed about why you were to take it, how to take it, and any side effects that might occur? If you have affirmative answers to all of these questions, you may well have found the "right" doctor for you.

121. WHAT SHOULD I HAVE READY FOR THE DOCTOR INTERVIEW?
You need to do some work to get the most out of the interview, while taking the smallest amount of the physician's time that you can. You should prepare a comprehensive family medical history, know the reason for your visit, and the symptoms you are experiencing. If the visit is specifically about menopause, learn as much about your symptoms as possible so that your exchange of information can be as valuable as possible to you both. Remember it is not reasonable to expect that a physician can answer a lifetime of questions in a single visit, so if

this is to be your doctor, expect that shared information, like trust, builds over a period of time.

I cannot stress too strongly the importance of finding the right doctor. Ideally, this should be a relationship from which you derive security and comfort. This should be akin to how you relate to other important people and professionals in your life—your lawyer, banker, accountant, or your religious leader. You need to find a doctor you can work with and believe in! No, please don't think I am creating an analogy of doctors as gods. Physicians have no interest in that role. As a matter of fact, I read recently somewhere that the notion of thinking of physicians as gods will be dispelled just as soon as we get up off our knees and take control of ourselves.

To empower you to take charge, I offer the following:

Twelve Tips for Making the Most of Your Physician Visit

1. When making your appointment, try to let the receptionist know how much time you may need, if you know.
2. Know the financial requirements of that office and have your insurance information handy.
3. If you are concerned by a long wait in the doctor's office, call ahead to see how the appointment schedule is running and whether you should arrive at a later time.

4. Don't be late (unless item 3 applies).

5. Let the receptionist know you are there, as soon as you arrive.

6. Bring a written list of your medical symptoms or complaints.

7. Bring a written list of your questions or the subjects you need to cover.

8. Make sure you fully understand each of the doctor's answers. If you don't, ask for further explanation at *that* time.

9. Eliminate small talk or chitchat with the doctor, whenever possible. The physician has a limited amount of time and wishes to render the best possible care. Getting to the point saves both of you valuable time.

10. Present yourself as confident and your problem as serious.

11. Before you leave, fully understand the need for further tests, therapy, or medication if indicated and the side effects, if any.

12. If you are unhappy with your doctor and cannot resolve your problems through dialogue, then don't hesitate to seek medical care elsewhere.

Remember: A satisfactory patient/physician partnership requires mutual respect!

CHAPTER 12

How Do I Go from
Ho-hum Back to
Hot Sex?

As I explained earlier in this book, there is a surprising scarcity of questions regarding sex during the question-and-answer period of the programs. Some things are hard to change, and I guess that it is still difficult for many women and men to frame questions about sexual desires and difficulties. However, an analysis of the anonymous questionnaires obtained at these programs reveals that more than half of the women list sexual difficulties as one of the issues they have come to the programs to learn more about. In contrast, the Gallup study showed that approximately 80 percent of the women reported no concerns about the effects of menopause on their sexual relationships. Yet, some women do come to me after the program with some version of the same complaint. So let's talk about having hot sex instead of hot flashes.

Women confide, "I wonder who arranged this situation. Finally, I don't have to worry about pregnancy. I no longer need to use birth control. I have a period that is so short that I don't consider it an impediment to sexual activity for more than a couple of days. Now that sexual freedom has arrived for the first time in my life, *I'm just not interested*."

One woman faulted the popular "use it or lose it" philosophy of sex, saying, "There's no point using 'it' if you feel nothing." Others comment, "I would use 'it' more if it didn't hurt so badly." Or, how about this familiar complaint: "My husband and I were always a sex-is-best-in-the-morning couple. Now I sting so badly when we're done, I can barely get up to go to work. No more morning sex for me!"

One woman added, "Another thing that's hard to explain is that when we were young and I was aroused and wet, he was hard and strong. Now, I'm dry and tight and he's often too soft to penetrate me." Or, "Now that it takes me so long to get aroused, he can't wait long enough to get me to orgasm. I love my guy, but I feel like a victim."

During these confidences, most women tell me that they care deeply about the quality of their sex lives. I believe that this discrepancy between the questionnaire responses and the Gallup study reflects the fact that the women and men voluntarily attended the programs in order to learn more about menopause. Given this high level of motivation, they may be more willing to search their hearts and minds on the subject of sex as it relates to the menopause. They also have anonymity in filling out the lengthy questionnaires while there. In contrast, the Gallup survey, conducted randomly by telephone, has many information-limiting biases, such as many people's basic discomfort of talking about sex on the telephone and the fear of being identified.

Nevertheless, there is much interesting information to be learned from the Gallup study. Approximately half of the menopausal and postmenopausal women in that survey said that they had looked forward to menopause because it brought with it freedom from worrying about unwanted pregnancies. A full 40 percent of the premenopausal women felt that way as well. Just 10 percent of those women believed that "after menopause a woman's ability to enjoy sex is greatly reduced." Interestingly, the men interviewed by telephone weren't nearly so eager in anticipating their wives' menopause. While admitting that they have only limited knowledge of the whole process of menopause, 45 percent of the partners of menopausal and postmenopausal women nonetheless expressed concern about the negative effects of menopause upon their sex lives.

Attitude and expectation play key roles when it comes to women and their sex lives at menopause and later in life. The Gallup survey heavily underscored the fact that the small number of women who were worried about their sexual relationships—who believed that menopause would change their life-styles no matter what they did, who thought that nothing could be done to ameliorate their physical symptoms, and who had decided that their ability to enjoy sex would inevitably diminish—reported that their interest in sex had decreased. Quite the opposite was true for women who had higher and better expectations about their continued good life-styles and sex lives and who had found a doctor with whom they felt they could comfortably discuss all of those issues. This type of self-fulfilling prophecy held true for the men as well.

Society has long equated youth and physical attractiveness with desirability. In the Gallup survey, physical attractiveness and the effects of their own aging appeared to concern more men than women, which is interesting. Perhaps you would care to explore this issue further in your own relationship. That question alone could stimulate some first-rate communication between you and your mate. Further, if as a result of your discussions, you and your partner agree on shaping and sharpening up for one another, this sharing of information and feelings would have encouraged self-improvement and self-nurturing activities for you both. Some of these activities will be described in the chapters on life-styles, Chapters 13 and 14.

Most of the postmenopausal women I've talked to who expressed disappointment in their sex lives wanted to continue to be aroused by things that usually aroused them, and they hoped to continue to be able to conjure up the fantasies that had worked to arouse them in the past. Women who were accustomed to experiencing orgasm expected always to complete sex that way. They wanted to know how to jump-start their sexual desire again.

I know that this is somewhat contradictory data, but by using it, I have tried simply to illustrate the fact that not only is human sexuality a complex subject, but that at midlife the attitudes and expectations about sex are complex as well. There are apparent contradictions in what people say they are willing to settle for in terms of their sex lives and what they really want. For example, some women and men may indicate that a ho-hum midlife sex life is okay with them, yet in their heart-of-hearts they desire a passionate one.

Today, there appears to be an upswing of medical interest in helping us achieve what we want sexually. We need to seek out sound information about having hot sex in our fifties, sixties, and beyond; we need to find a physician with whom we can communicate honestly and comfortably about sexual matters; and we need open lines of communication with our partners. Motivating ourselves to find and use these necessary resources can significantly affect what happens to our sex lives.

The history of female human sexuality research is pitifully short, beginning in earnest only in 1953 with the publication of *Sexual Behavior in the Human Female*, the landmark survey by Alfred Kinsey. Sexuality research moved forward again, in 1966, with the work of William Masters and Virginia Johnson, who, in their book *Human Sexual Response*, describe how changes in one's physical anatomy can alter sexual response. (I had the privilege of dining with Dr. Masters when he came to Cleveland to speak at a seminar. He appeared to be a man so committed to his work that, although everyone at the table vowed not to discuss sexual response at dinner, it was the subject he preferred.)

The more recent work of leading authority Dr. Phillip Sarrel at Yale University indicates that both males and females have a diminution of sexual interest and response as they age; that one person's disinterest or dysfunction affects the other; and that a previously healthy sexual relationship has a good chance of surviving menopause and aging and may even thrive as a result. That is very encouraging information!

In terms of your menopausal medical care, it is crucial that your physician treats you as a whole woman, taking all of your attitudes, expectations, and fears into consideration in determining the course of your care and treatment. You should feel that you can communicate freely about your sexual needs and that your doctor is willing to work cooperatively toward solutions for your sexual concerns and problems. Don't give up your good sex life!

A close friend of mine told me the following story. "I was only in my forties when I began to think I was too old for sex. What I now know is that vaginal dryness made sex so uncomfortable, I thought I would just give it up. However, my husband didn't agree and couldn't understand why I would not discuss this problem with my doctor. I just couldn't. Finally, my sister told me about a new vaginal moisturizer that she was using. I ran out and bought some. What a difference! I'm back to my old self."

122. WHY DOES SEXUAL DESIRE SOMETIMES DISSIPATE? IS THERE ANYTHING A WOMAN CAN DO TO RECAPTURE IT? This is one of the big pluses of ERT or HRT. Studies show that 90 percent of the postmenopausal women surveyed returned to their own normal degree of desire once they began hormone replacement. There are also non-hormonal ways to stimulate desire: There are vaginal contractions that we can do to keep the muscles of the vagina toned; water-soluble vaginal moisturizers to eliminate dryness and offer more flexibility to the vaginal walls; and there are ways to enhance physical appearance (following the look-better/feel-better/feel-more-sexual philosophy). Claims also are made for the use of certain herbs, for visualization and fantasy therapies, and for erotic books and videotapes. Whether or not you're on ERT, try varying your sexual experiences now that you have more time or space to do so. Have sex in the living room in front of the fire, if your nest is empty. Check into a motel or hotel in your own hometown for an away-from-home sexual escapade. Plan to meet each other at home in the middle of the afternoon after making up excuses for your absences from wherever you are sup-

posed to be. Be inventive, be creative, become a sexual gourmet. Even though every sexual meal may not be perfect, you will maintain your interest and that of your partner in finding exquisite sexual experiences.

123. How have I changed physically to make sex so darned uncomfortable?

Many systems, tissues, and organs in your body depend on estrogen for nourishment. Your pelvic structures have that same dependency; therefore, sexual function is affected by the decline in estrogen. Dr. Phillip Sarrel explained it this way in a supplement to the April 1990 issue of *Obstetrics and Gynecology*, a professional journal.

> . . . *cell growth and multiplication decline, since estrogen acts on cells as a growth stimulant. Loss of cells in the vaginal lining leads to thinning of the tissue and increased susceptibility to irritation and tears during intercourse. . . . Touch perception declines, which may make a woman less sensitive to tactile sensations that are an important part of sexual stimulation. . . . Blood flow to the genitals, and possibly to the heart and brain, declines, which may decrease engorgement of vaginal and other tissues associated with sexual stimulation. Estrogens increase arterial blood flow.*

The above explains how, when your supply of estrogen has dwindled, the lining of your vaginal canal thins; thus the entrance of the penis can hurt. You may feel a burning sensation or the sense of actually being torn. ERT helps by replacing missing estrogen to the pelvis, as well as to the rest of the body. This may also be a good time to try some new sexual positions that are more comfortable and stress your tissues less. For example, one woman told me that she solved her problem by always being on top and therefore in control of the thrust of the penetration. She eliminated her fear of being hurt and solved her sexual problem. Sex became fun again!

124. Why can't I feel aroused the way I used to? My husband used to touch me in certain ways in special places and I tingled all over. Now, I just feel vaguely annoyed.

Think about the various fabrics of your clothing. Does silk and satin still feel luxuriously smooth? Does mohair still feel warm and cozy, or does it itch like crazy? Did you once like the feeling of fur or plush on your bare skin? Do you still? Do freshly starched sheets feel crisp and clean and wonderful, or do they irritate you now? These tactile changes may be the result of changes in skin sensitivity that occur with estrogen loss to the sensitive nerve endings of the skin. This may be one reason why you no longer get a sexual surge from a formerly favorite kind of caress. The problem is complicated, but it is not insurmountable. Skin sensitivity usually returns with ERT. If you can't or don't want to take estrogen, there are still ways around this sensitivity issue. Experiment. It may just take you time to find different routes for arousal. Don't be afraid to seek sexual arousal through erotic literature or videotapes, in the privacy of your own bedroom. Try using a vibrator alone or in conjunction with your partner, and experiment with the degree and location of vibration and lubrication that stimulates you. There is no "norm" to be concerned about. You should care only about whatever works best for you and, of course, for your partner.

125. EVERY ONCE IN A WHILE WHEN I GET INTO BED, I GET THIS CREEPY, CRAWLY SENSATION. IT'S AS IF ANTS ARE CRAWLING ON MY SKIN, BUT THERE'S NOTHING ON ME. I HAVE SCRATCHED UNTIL I'VE BLED. DOES THIS HAVE TO DO WITH SEXUAL AVOIDANCE OR COULD THIS BE PART OF MENO-PAUSE?

What you are describing is one of the rarer menopausal symptoms and it is not related to sex or to being in bed. It is a condition called formication and your description of it is apt. (The word *formication* actually comes from the Latin, *formica*, and means "ant.") It is one of the stranger symptoms of menopause and is an infrequent occurrence that can be gotten rid of with ERT. This is another important symptom to understand and explain to your mate, so that your reaction is not taken personally.

126. WHY DO SO MANY DOCTORS TALK ABOUT AND WRITE ABOUT THE "USE IT OR LOSE IT" PHILOSOPHY OF SEXUALITY AFTER MENOPAUSE?

They mean that continuing with sex will enable us to

continue to be able to have sex. Perhaps menopause, with its freedom from the risk of pregnancy, is the time to search for the great romance of your life if you are single—provided, of course, that you practice safe, protected sex.

One woman in New Orleans told me that she was asked for the first time ever whether her sex life was satisfactory. Only this year, at the age of sixty-three, was she asked that question by her new internist. She had never been asked that question by her gynecologist in all those prior years! This is why I again urge you to find a doctor who treats the whole you and in whom you can confide about your sexual needs and any concerns about your lack of sexual interest. Your doctor may have suggestions which range from a consideration of ERT and improved diet and exercise plans to finding a support group or counselor to work on your attitude and behavior. Be prepared to ask enough questions to assure yourself that your doctor takes your sexual problem seriously, understands that sex is important at any age, and is not just trying to avoid or get rid of a problem that he or she is not comfortable dealing with. Most physicians who specialize in treating menopausal women suggest that you "use it or lose it" because sexual activity is an important consideration in keeping your sexual organs functioning well.

127. HOW DO YOU SUGGEST I "USE IT," IF I DON'T HAVE A PARTNER?

Masturbation can be a satisfying and a satisfactory choice. In Victorian times, it was thought that masturbation could ruin your mental and physical health. Remember the old saying, "Don't do that, or you'll go blind"? Masturbation was misunderstood then, and to a great extent it is still misunderstood today. Many of us still believe the old taboos. But experts assure us that masturbation is a healthy means of handling sexual needs and expression when there is no partner available, or when you don't want one. It is a personal, pleasurable way to release pent-up sexual energy, and it can enhance sleep by reducing tension. Masturbation works best when you have a good fantasy life. There are many books on the market that describe masturbation techniques. A good lubricant and a hand-held vibrator are mentioned as a favored technique by many women. Others enjoy a hand-held massaging showerhead. Through masturbation you can learn about your own body and about what pleases and excites you, either for yourself, or to teach your partner how to make sex better for you—perhaps for you both.

128. I HAVE BEEN ON ESTROGEN REPLACEMENT THERAPY FOR SEVERAL YEARS, AND ALTHOUGH IT SEEMS THAT MY SYMPTOMS OF MENOPAUSE ARE ABSENT, I *STILL* FEEL NO DRIVING INTEREST IN SEX. WHY?

The answer may not be far off. Scientists are working to understand fully the effect of adding androgen, the male hormone, to ERT or to combined hormone therapy, HRT. The logic behind this is that the male hormone appears to govern our sex drive. The desired result has been found with this new treatment, as libido increases in many women. Some physicians are prescribing androgen as a desire enhancer, with good results. This work is fairly new and, like all information pertinent to your health care, should be discussed with your physician. The introduction of the male hormones must be very carefully administered because although they can stimulate sexual desire, they may promote facial hair growth and other male characteristics. (Theoretically, androgens may also compromise some of the possible cardiovascular benefits of HRT). Proceed with caution! For women who are not on ERT, another theory suggests that women's sex drive may increase after menopause because our level of testosterone, a male hormone that we all have, becomes dominant as estrogen levels decrease.

129. MY HUSBAND IS WORRIED ABOUT WHAT IS HAPPENING TO OUR MIDLIFE SEX LIFE. HOW CAN I HELP HIM TO UNDERSTAND?

Try to get him to accompany you to a consumer education program on the subject. Many medical centers are holding such events for exactly that purpose. Our programs used to be attended solely by women, but that has changed over the last three years. Now men make up at least 20 percent of our audience. Stimulate discussion by sharing interesting articles and books that you read on the subject. Last, and very important, set up a consultation for you *and* your husband with your doctor, with sexual concerns as part of the agenda.

130. ARE THERE ANY PELVIC EXERCISES THAT CAN ENHANCE SEXUAL PLEASURE?

Some women suggest that pulling in on, or tensing, the muscles in the pelvis, buttocks, and thighs during sex can add to more intense feeling of sexual stimulation. Kegel exercises, described on page 149, can help as well.

131. HOW DO I BECOME MORE SEXUALLY FIT?

A sense of sexual fitness comes from the mind and the body. They are so interrelated that sexual arousal can begin with something we read or watch and quickly create changes in our anatomy. Women may become vaginally lubricated and their nipples may become erect. In men, the penis may become erect. It follows then, that exercise, if it translates for you into a feeling of pride about your body and serves as an elevator of mood, can make you feel mores sexually fit as well. A body weight at which we are comfortable and a shape in which we take pride are also sexual enhancers for some. For example, a survey on sexual fitness of more than five thousand readers by *American Health Magazine* in its December 1991 issue indicated that men are more interested in sex when they are feeling fit and "in shape," and women feel sexier when they weigh less.

Questions and answers concerning exercise and diet are in Chapter 13.

The sex education book that I hid under my mattress, beginning with its publication in 1973, was *The Joy of Sex* by Dr. Alex Comfort. It was there that I learned about the sexual joy and abandonment to be found in many positions practiced by many cultures. The millions of copies of *Joy* sold attest to the fact that I was not alone in my pursuit of knowledge and technique. In 1991, Dr. Comfort offered us a new book, *The New Joy of Sex: A Gourmet Guide to Lovemaking for the Nineties*. It's quite different. There is more regard for the equality of enjoyment for women as well as greater interest in the health and fitness of the depicted lovers, whose bodies are now streamlined and in living color. In addition, in Comfort's new book, the discussion of joy and pleasure is combined with guidelines for safer sex.

The sex instinct is one of the three or four prime movers of all that we do and are and dream, both individually and collectively.

Philip Wylie
Generation of Vipers,
1942

CHAPTER 13

What Can I Do to Feel and Look My Best?

Life-style-change Rx: There are innumerable activities that you can pursue on your own behalf to make you feel better and look better. However, before you engage in new exercises, diets, or vitamin regimens, consult your doctor. What works for one woman simply may not be good for another. So check it out!

This morning I got up, brushed my teeth, weighed myself, and had my breakfast of one half an orange, fiber-loaded bran cereal with skim milk, and a steaming mug of coffee with more skim milk. Then I had a fourteen-ounce glass of water and took my morning all-purpose multivitamin, Vitamin B-complex, 1,500 mg of timed-release Vitamin C, 400 IU (International Units) of Vitamin E, plus 25,000 IU of beta-carotene, and 50 micrograms (mcg) of selenium—and surveyed my morning regimen. It made sense for me. I feel well. In fact, I feel terrific.

I put on my workout clothes with a sweatsuit and got ready for my daily workout, which I usually schedule about two hours after breakfast. Then I went to work, taking advantage of my morning breakfast's surge of energy to get some cerebral work done in my home office. Just as my brain and bladder were about to let me know that it was time to take a break, I realized that it was exercise time. That is how many of my days begin.

I relish letting off some steam and working up a sweat while depending upon the answering machine to keep track of my telephone calls. It is Monday and an aerobics day for me. I do an aerobic workout on Monday, Wednesday, and Friday, each session lasting from twenty to thirty minutes. Luckily, I can choose my aerobic pursuit, which for me staves off the boredom of repetition—treadmill, recumbent bicycle, cross-country ski machine—or the simple variety offered by nature: a walk out-of-doors on a beautiful day. Before beginning any form of aerobic or weight-lifting exercise, I spend ten minutes on stretch and flexibility exercises. After my aerobics session, I do cool-down stretches for another ten minutes. From beginning to end, I drink another large glass of water or more. All in all, my aerobic exercise takes less than one hour of my time.

On Tuesdays, Thursdays, and Saturdays, I lift weights using both free weights and machines. These workouts, too, are preceded and followed by ten minutes of flex and stretch. I take off on Sunday or use it to make up for another day when I couldn't fit in exercise. Make no mistake, my program is not perfect by any means. Some days I goof off, but it is enough of a regimen to make sure I get enough exercise or, at least, to feel cheated when I don't. Still, nobody's perfect!

I made up my schedule after checking with my doctor. It works for me now, because I work at home. The point of sharing it with you is to encourage you to look at your days and fix certain "for me" times in them. When I first went to work as vice president of Cleveland's Mt. Sinai Medical Center, I used to exercise in the evenings, because our daughters needed all of the morning before-work, before-school time. When they were grown and gone, I began to exercise on my lunch hour. Now that I can schedule my days to my own liking, I have changed to the routine that I described earlier. That is what works best for me now.

I offer all this personal information because when I lecture, women almost always ask what form of exercise I pursue, what diet I like best, and what I do to maintain young-looking skin. In this chapter, I will try to answer these and other questions not only by sharing information on what I do, but by sharing what I have learned we should all do, provided that medically it is good for us as individuals.

A Consumers' New Attitude Survey commissioned by the Magazine Publishers of America and reported in the October 7, 1991 issue of *The New York Times* clearly demonstrated how the desires and needs of people differ from their actual behaviors when it comes to nutrition and exercise. This survey of twenty-five hundred people in two-hour interviews by the annual Yankelovich Monitor (a survey conducted by Yankelovich Clancy Shulman, a marketing and social research group), showed that today more people than ever before *say* that they are concerned about maintaining the right weight, getting enough exercise, eating a balanced diet, cutting back on high fat and high cholesterol foods, watching their blood pressure, and getting enough rest. Their behaviors, however, don't necessarily bear out their desires. Although fewer persons say that they eat what they want and don't worry about nutrition when they eat out, the percentages of both women and men who say that they are eating healthier diets in 1991 have dropped.

That means to me that we have to rededicate ourselves to the behaviors and practices that will keep us fit. I am so grateful that the heyday of the Twiggy body has ended and that we are permitted to have strong, well-shaped, well-rounded bodies that let us look like women, not stick figures. Today's models emanate a sense of control, strength, vitality, and good health. You could expect to see them on the tennis court or in the workout room just as easily as in the sauna or on the massage table. They may pamper themselves, but they also pump iron. One of my new idols is Linda Hamilton, female star in the movie *Terminator 2: Judgment Day*. Although the movie was pure science fiction, she was portrayed as a compassionate care-giver: feminine, yet lithe, well-muscled, and super-strong. She conveyed the image of a woman who can take care of herself—and in that movie, she does. I wonder what she eats!

One of the most-often-asked life-style questions at the seminars is about diet. Let's tackle that one.

132. WHAT IS A GOOD DIET FOR THIS TIME IN MY LIFE? That's a good question, and women usually write it on their personal questionnaire in just that way. When I am asked that question, I respond by explaining that a good diet for midlife is probably the diet we should have been following lifelong. Most experts agree that good health and high energy are influenced by our diet. Our diets should be directed at lowering fat and increasing fiber intake. However, it is not my intention in this book to offer a diet program—either seven-day, fourteen-day, or any number of days or weeks or months. In the course of my life, I have probably tried every crash diet ever invented, always in search of that five-to-ten-pound quick loss. I have lost and regained those pounds any number of times and only my constant exercise programs have kept me from yo-yoing myself into a body composed largely of fat. So no quick diet, no diet plan, and no menus or recipes are included here. Sensible, long-term habits are the name of the game.

133. WHAT IS THE BEST WAY TO TAKE OFF EXCESS WEIGHT?

I want to stress the idea of reducing fat in the body by greatly reducing the fat in our diets. That's what I've finally done and it works. It's based on the knowledge that fat makes fat! Rather than reducing your overall daily food consumption and rather than counting calories, cut your fat intake. I try to keep mine between twenty and thirty grams per day. No, don't cut out *all* fat or your skin, hair, and nails may suffer first; then your body systems will follow suit. Fat is also important in providing us insulation from the cold as well as providing a storehouse of reserve energy to draw upon when we are under duress. Ounce-for-ounce, protein and carbohydrates contain fewer than half the calories of fat. But it is the fat that plants itself fastest on your stomach, hips, and thighs, precisely because it packs those extra calories and less nutrition. When you think of *fat-free*, think of most fruits and vegetables, and of grains and beans. There are some exceptions, however—like the avocado, the olive, and the coconut—that are largely fat.

134. WHAT CAN I DO TO CUT OUT FAT WHEREVER POSSIBLE?

Remember that most of your fat grams will come in dairy products and meats. No, don't eliminate them; alter them. For example, exchange whole milk for skim milk; ice cream for nonfat frozen yogurt; and limit your portions of fish, poultry, and meat to approximately six ounces per day, divided however you wish. (In addition, start being a label reader when it comes to buying packaged foods and a careful menu reader when it comes to eating out. Some restaurants, even fast food ones, are making an effort to offer low-fat foods, but if they don't, do not hesitate to ask how a dish is prepared or to state how you wish to have it done. I often will see a pasta with vegetables in cream sauce on a menu and another one of cheese- or meat-filled tortellini in a tomato/basil sauce. I have never been refused when I order the pasta with vegetables but specify that I want it with tomato/basil sauce. That's just one example of how to customize your meal at a restaurant. Don't feel uncomfortable about asking for fat-free or low-fat alterations; restaurant kitchens are becoming accustomed to such requests. According to American Heart Association dietary guidelines, your daily fat intake should be less than 30 percent of your total calories.

For more information about lowering fat consumption, I've listed some books that worked for me in the Appendix.

135. WHAT CAN I DO ABOUT CONSTIPATION?
Constipation is a problem for many women. Therefore, the second nutritional change that can be important is adding fiber to your diet. Bran cereal with extra fiber for breakfast every day works for me. I also seek out other good sources of fiber, particularly broccoli and strawberries, because I like them. Most plant foods contain fiber,

which aids in elimination and assists with other lower digestive tract problems as well. Good natural sources of fiber include fruits and vegetables, whole grain breads and cereals, and dried beans and peas. The latest nutritional information from the National Cancer Institute reveals that we should include at least five helpings of fruits and vegetables each day. Drinking sufficient amounts of water, about eight large glasses daily, also helps to flush out waste. Exercise also helps to relieve constipation.

136. How much fiber should I eat each day?
The American Institute for Cancer Research recommends between twenty and thirty-five grams of fiber for women per day. I caution you, however, if you have been consuming far below that amount, to increase your fiber intake slowly over time, so that you do not become bloated and uncomfortable. In addition to relieving constipation and keeping the digestive tract in good working order, fiber also helps to prevent us from absorbing a small amount of the fat that we eat. It just whisks it away. Don't get carried away with that idea, however, because too much fiber can also carry away other vitamins and minerals that we need.

137. How much calcium do I need each day?
Probably more than you are getting. The average woman consumes approximately 500 milligrams daily in her diet. As of this writing, the RDA indicates that we should

consume 800 milligrams. So we're coming up short to begin with. Nutritionists suggest that in the premenopause we should take in 1,000 milligrams and that we need 1,500 milligrams after menopause, if we are not on ERT. If we are, then 1,000 milligrams remains enough. Calcium is our bone builder and preserver. Good sources of calcium-rich foods and calcium supplements are listed in Chapter 8 where osteoporosis is discussed. The latest calcium information indicates that calcium works best when taken with magnesium.

138. CAN YOU GIVE ME SOME TIPS ON HOW TO MAKE FOOD TASTE GOOD FOR THE REST OF THE FAMILY ONCE I BEGIN TO ELIMINATE FATS, SALTS, AND SPICES? I HATE PREPARING MULTIPLE MENUS.

I've learned to enhance the flavor of foods in a number of ways, while eliminating the fats and the heavy spices that may bring on hot flashes. I use nonfat yogurt in place of sour cream; fresh herbs (which I love to grow and harvest as needed), minced garlic, minced onion, and cilantro to replace salt; flavored vinegars, orange juice, tomato juice, honey, black pepper, and lemon juice; mustard and light tahini and light soy sauces ("light" or "lite" versions of these brands contain less salt) for flavorings and in salad dressings and marinades. There exist many other good flavor substitutes as well. With a little research and trial-and-error, you can create your own style of healthful cooking.

139. WHY DO YOU STILL TAKE VITAMIN E DAILY?

I started taking Vitamin E for night sweats. I continue to take it now because medical studies have shown that 400 IU (International Units) of Vitamin E, taken daily for more than two years, helps to prevent heart disease. Vitamin E dilates the capillaries and thus improves circulation. It also helps to prevent blood clots and may even dissolve existing clots. I take it to help protect myself from these risks. Why not?

140. WHAT DO BETA-CAROTENE AND SELENIUM DO?

A recent study demonstrated that taking 50 milligrams of beta-carotene a day reduced the risk of heart attack and stroke by one half. Selenium was shown also to prevent heart disease.

141. IS WALKING REALLY GOOD ENOUGH EXERCISE FOR ME AT THIS TIME OF LIFE?

Walking is the most popular exercise in America, according to a survey done by the National Sporting Goods Association in 1989. There are now more than sixty-five million exercise walkers. In general, many doctors suggest walking for thirty to forty-five minutes a day at a pace of three to three and a half miles per hour for beneficial results. However, your walking program should be designed for you, taking into consideration any medical problems you may have, or any fitness goal you want to achieve. Walking is now considered a good choice of weight-bearing exercise at any age and it can be done year-round in nearly every environment.

142. IS WEIGHT TRAINING REALLY SAFE FOR SOMEONE MY AGE? I'M FIFTY-THREE AND HAVE NEVER REALLY EXERCISED IN MY LIFE.
It can be, if your physician agrees. Today, total-body fitness for healthy adults includes workouts with weights to make our muscles strong and to keep our bones dense and help us to ward off osteoporosis. In fact, the American College of Sports Medicine has acknowledged the importance of resistance training and recommends that healthy adults perform weight workouts twice a week. Their suggestion is for us to complete eight to twelve repetitions of eight to ten exercises at each workout session. Another plus of weight workouts is that the resulting increase in muscle mass speeds up our metabolism, causing us to burn more calories.

143. HOW OFTEN SHOULD I EXERCISE DURING MENOPAUSE?
Exercise should be a regular activity, performed at least three days a week. Aerobic exercise should be continuous for twenty to thirty minutes at 60 to 75 percent of your maximum heart rate, which you find by subtracting your age from 220 and then taking the appropriate percentage of that number as your maximum. For example, if you are forty-five years old and you subtract 45 from 220, the remainder is 175. Then if your doctor agrees, you take

70 percent of that number and learn that the maximum number of heartbeats per minute that you should work out at is 122. Never exceed that number and you are working in a safe range for you. Workouts with weights are also suggested and often these can be done on alternate days.

144. HOW DOES MENOPAUSE AFFECT WEIGHT LOSS AND KEEPING WEIGHT OFF?

The number of calories we women need declines as we age. It is estimated that the number declines between 2 and 8 percent for every decade after our twentieth birthday. Regular exercise can help to make up for that decline by speeding up our metabolism so that we burn more calories. Also, as we age, our percentage of body fat increases and our lean muscle mass decreases. A sensible low-fat diet and exercise can help to reverse that change.

145. HOW CAN I HELP KEEP MY SKIN FROM LOOKING OLD?

Our skin naturally changes as we age, becoming thinner and dryer and less elastic. However, there are many things that we can do to offset this process. Most important, keep your skin out of the sun as much as possible, especially between the hours of 10:00 A.M. to 2:00 P.M., when sunlight is the strongest. When you are in the sun, use a sunscreen of at least 15 SPF and reapply it often. Don't smoke! Smoking accelerates the natural process of aging and may even add wrinkles around your mouth from the repeated puckering-up action of inhaling smoke. Exercise speeds blood circulation, which feeds the skin. A well-balanced diet with high levels of Vitamins A and E maintains healthy skin—good food sources of Vitamins A and E are carrots, squash, peppers, leafy green vegetables, yams, and vegetable oils. Drinking your eight glasses of water a day also nourishes the skin. Proper cleansing and moisturizing of the skin is vital. I use a gentle cleanser on my face and neck followed by a stimulating toner and moisturizer each morning and again before bed. I reapply the moisturizer several other times during the day, especially if my face is subjected to sun, wind, or extreme cold.

146. How do I eliminate the negative stresses in my life?

Stress is a part of our daily life. In modern usage, we tend to define stress as extraordinary demands put upon us to which our bodies and our minds respond. Years ago, when I reviewed lay medical books for Cleveland's newspaper, *The Plain Dealer,* I wrote what I had learned from one book that "battle fatigue, shell shock, and stress have the same underlying symptoms: fatigue, irritability, impaired judgment, and loss of confidence." I still believe that is true and that the severity of our symptoms will depend on how prolonged our struggle. You will note that I always speak of reducing *negative* stresses in our lives. That is because certain stresses are positive, such as those that motivate us toward reaching our goals. It may be the stresses surrounding menopause that actually cause some of our psychological changes, such as for-getfulness, minor depression, or anxiety. Environmental stresses that occur around the age of menopause can be monumental to some of us when we couple landmark physical changes with the loss of children from our nests, or of parents through illness or death. When I say that it is important to try to remove negative stresses from our lives at this time, I mean that it's advisable to unload as many unnecessary burdens as we can and learn to cope better with those we must carry.

147. How do you de-stress?

I often make a list of what is bothering me and try to eliminate what I can. For example, when I look at my list and see that I am overcommitted to home, job, or family, I know that I have to make some decisions about unloading something. If it's work, I may need to extend a deadline or drop a project. If it's home, I may decide not to cook this week until I catch up on other tasks overloading me. If it's my mother's illness that is troubling me, I may realize that I need to talk that stress over with a therapist. For me, the worst stress comes from not dealing directly with stress. So I suggest that you make a list of what is stressing you, study it, determine how you can cope best with each item, and discard whatever you can. If you need outside help to diminish stress, go for it. Remember that the days of trying to be "supermom" or "superwoman" should be behind us.

148. IS THERE ANY SIMPLE WAY TO OBTAIN RELIEF FROM STRESS?

If in making your list of stressors as described above, you can identify situations that are frequent causes of stress for you and can determine an identifiable source of that stress, put a red dot on it using a washable red magic marker or a sticker. For example, if you frequently are running late and that is stressful for you, put a red dot on the face of the clock you most frequently check and one on your watch face as well. If other drivers honking at you when you're sitting in traffic stresses you, put a small red dot on your rearview and sideview mirrors. Locate as many sources of stress in your life as you can and red dot them (only inanimate objects, please!). Each time you see your red dots, take three slow, deep breaths. This will cause an automatic relaxation response and enable you to switch gears—from high to low. Within thirty days, you should begin to notice that the things that always bothered you are less bothersome. Keep up the good work. Every once in a while, reevaluate your red dots and eliminate those you no longer need or add others where you've identified stress locations.

There are many other subjects that could have been included in this chapter on life-style changes, but I have confined my answers to those questions that are actually raised at the educational programs at which I speak. I believe that when we work to incorporate healthy diets, appropriate exercise programs, and stress reduction techniques into our life-styles, we are on our way toward the first-rate second-half of the life that we desire and deserve.

> *Grow old along with me!*
> *The best is yet to be,*
> *The last of life, for which the first*
> *was made.*

Robert Browning
"Rabbi Ben Ezra," *Dramatis Personae*, 1864

CHAPTER 14

What Else Can I Do to Enjoy the Second Half of My Life?

Live your life to the fullest extent that you can. Carve out as many hours, days, and weeks as you can that aren't otherwise committed and call them your own. There is inestimable value in the wisdom you have gained throughout your years of living. Now is the time to use it on your own behalf.

Two weeks before I said good-bye to 1991 forever, I read a fascinating magazine article about Gloria Steinem—a woman whose mission I had always admired along with her beauty. The *Vanity Fair* article told of her life, including the sadness as well as the accomplishments. It told of her inability to use the strength that was gained from her various life tragedies to benefit herself—although, perhaps, she had been driven by those tragic experiences to benefit the women's movement. Until now. Now postmenopausal and still beautiful, she has come to understand the lonely child within herself.

The message for us comes at the end of the article when author Leslie Bennets poses a question about what the future will hold and Steinem responds: "I feel as if now I know what I want to say, I should begin it."

We all should begin our futures, with our own goals in mind. We exist, still lovely, and in our large numbers and with our energetic drive we are capable of creating change—of influencing an election, of having an impact on issues of sexual harassment, world hunger, saving the planet, homelessness, and child abuse, to name a few. We've all had our scares and shake-ups—even Gloria Steinem had her cancer scare and lumpectomy—but we should have emerged from these negative experiences even stronger and more committed to doing what we want to do. We've earned the right to quit caring so much about what other people think of us. It's passé. Who cares?

There are so many of us at midlife. Just look around. Or look in or at the media: There's Barbara Walters, Joan Rivers, Dr. Bernadine Healy, Gloria Steinem, Gail Sheehy, Elizabeth Taylor, Jane Fonda, Diana Ross, Raquel Welch, Olympia Dukakis, Shirley Jones, Shirley Maclaine, Pat Schroeder, Geraldine Ferraro, and so many, many more. I could go on and on and you could add to that list. It is not the length of the list, nor the prominence of the women, that matters, however. It is the fact that we are not invisible; we are not asexual; we are a presence to be considered and to be reckoned with. We have lived one half of our adult life—ages fifteen to fifty—before menopause. We have the other half—a full thirty-five more years—to live after it.

149. HOW SHOULD I APPROACH THE SECOND HALF OF MY ADULT LIFE?

You should approach the gift of life with vim, vigor, courage, confidence, and conviction! I know that there are changes in our bodies and in our appearances that some of us do not welcome. However, our rich experiences and our positive attitudes can more than compensate for them. I know that some women are genetically programmed to age more gracefully than others. I also know that all women can maximize their health and appearance to whatever degree they wish. Perhaps when you read my list of prominent women, you thought of their genes, their plastic surgeons, their personal trainers, their nutritionists, and their seemingly limitless ability to cast a wide cash net to capture everlasting youth. Not so, I say. Many on the list have suffered debilitating illnesses, divorce, widowhood, and loss of loved ones. Central casting has cast some of them out. Yet they have gone on to make a difference in some other medium for some other cause. They fight war, AIDS, illiteracy, poverty, hunger, and the glass ceiling that is still screwed down tightly over women's heads.

In the previous chapter, I discussed our universal

need for exercise and a healthful diet for our good health. Our good looks come along as a by-product. So whether we choose a cosmetic surgery lift, or to color our hair, or to exercise and eat right so we can maintain our former treasured dress size isn't the point. That's personal choice. It *is* important, though, that we do it for ourselves, not to be imposters of youth, but to be creators of what the mature woman can really look and feel like. Helen Hayes, whether or not she ever had a face-lift, retains her position as first lady of the theater; the late Diana Vreeland was always considered to be the first lady of fashion; the late Martha Graham was the doyenne of dance. Fulfillment is not about age; it is about attitude!

150. HOW CAN I LEARN TO LOVE THIS TIME OF MY LIFE? To me, loving adult life has to do with continuing to want to make a difference, whether for ourselves, our family, our friends, or our society in general. This half of our lives is about choices and about control. In the Yankelovich survey, three new categories headed the list when the respondents were asked what represented status in their lives. Number one was "being satisfied with your life"; number two was "being able to afford the things that are important"; and number three was "being in control of your life."

Choices abound. Opportunities for control exist. There are so many needy places that would welcome our attention, so many causes that our efforts could benefit. In cities and towns everywhere, there are hot lines ringing that need to be answered by caring and concerned volunteers. They are ringing at rape crisis centers, abuse centers, and at Alzheimers' caretaker information centers, and as the ranks of older persons in our society continues to swell, more lines will be ringing. There are spas that cater to our needs, cruise lines that extend special welcome to us, and educational vacations, such as the Elderhostel program, that bring us closer to people of our own age and interest. There are jobs open to us, consulting positions waiting for us, lonely children hoping for a grandmother-for-a-day visit, and nannys desperately needed by the dual-income young families of today. There are also beaches to stroll, shells to collect, birds to watch, gardens to grow, books to read, holidays to celebrate, lessons to learn, and love to give.

I have a retired friend who, on occasion, wistfully comments that truck drivers no longer whistle at her, but she is too busy relearning to play the piano to care for more than a moment now and then. I have another friend who is a concert pianist who still plays to the applause of thousands, but whose "real life" now delightfully revolves around her grandchildren.

Our attitude of belonging, of being needed, or of having something to say, is much more important than our thinning or fading hair or the slow encroachment of inches around our waistlines. No matter what the changes, our happiness and hopefulness are up to us.

I did not intend to make my list of opportunites for you complete. It need not be. You can add to that list your own needs, your family's needs, and the need of your friends and your community. And, of course, add your own heartfelt dreams and desires.

My goal in this book has been to show that there is good and rich land on the other side of our transition and that we need not fear or deny menopause, our inevitable rite of passage, but rather that we can welcome it just as each year we welcome the rites of spring.

Menopause is not our enemy. Aging is not our enemy. As I learned from Dr. Estelle Ramey, professor emeritus and senior physiologist at Georgetown Uniterity:

Our number one enemy is boredom—not the boredom of nothing to do, but the boredom of not doing anything that thrills or delights us.

A Final Note: Write to Me, Talk to Me

After finishing this book, I looked back at the large batch of questions that there wasn't room to include and realized that there are many other questions to be answered, perhaps another volume to be written. There are so many women with so many questions and I want to help find the answers for them, if I can.

So I have decided to try and keep the lines of communication open to my readers and to the women whom I meet at the programs so that we can all learn more about living a successful and happy second half of adult life from one another. Keeping in touch with you can make that possible. Obviously, no one—not even a physician—can handle medical problems by mail. Taking care of you medically requires that patient/physician partnership that I have been talking about throughout this book. But I know that there are other ways in which we can help each other by sharing ideas and life-style changes that have worked well for us and by keeping one another informed of what's going on in medical and in nonmedical treatments to help us.

So I've set up a mail box system to receive your letters with your questions, your suggestions, and your comments. I can't promise to answer each letter individually, but I will try, and you can count on receiving a card acknowledging receipt of your letter and letting you know that it's already been included in the pile of research data that I am seeking so that I can provide more consumer information to you as quickly as possible.

Please do not include requests for referrals to physicians in your area. The best source for that is the North American Menopause Society, the address for which is located in the Appendix (please write; do not call the Society). Look there, too, for other organizations, both national and in your locale, that can provide information or resource material for you, such as the National Osteoporosis Foundation. There are many others listed as well.

If you wish to share your thoughts and ideas with me in the hope of helping other women cope, please write to me at the following address:

Ruth S. Jacobowitz
c/o Croner Publications
10951 Sorrento Valley Road, Suite 1D
San Diego, CA 92121

A Final Note

Thank you for joining my consumer research project that is focused on continuing to learn and to share "what women really want to know."

Ruth S. Jacobowitz

Appendix A

Recommended Reading and References

Bailey, Covert. *The Fit-or-Fat Target Diet*. Boston: Houghton Mifflin, 1984.

Barbach, Lonnie, Ph.D., and David Geisinger, Ph.D. *Going the Distance*. New York: Doubleday, 1991.

Beard, Mary, M.D., and Lindsay Curtis, M.D. *Menopause and the Years Ahead*. Tucson: Fisher Books, 1988.

Benson, Herbert, M.D. *The Mind/Body Effect*. New York: Berkeley Books, 1979.

Blackburn, George L., M.D., Ph.D. *Prevention* magazine. "Eating Under Stress." June 1991, pp. 104–106.

Boston Women's Health Book Collective. *The New Our Bodies, Ourselves*. New York: Simon and Schuster, 1984.

Brecher, Edward M., and the Editors of Consumer Reports Books. *Love, Sex and Aging*. Mt. Vernon, NY: Consumers Union, 1984.

Brody, Jane E. *Jane Brody's Nutrition Book*. New York: Bantam Books, 1981.

Cobb, Janine O'Leary. *A Friend Indeed—for Women in the Prime of Life*, December 1991.

Comfort, Alex, M.D., D.Sc. *The New Joy of Sex: A Gourmet Guide to Lovemaking for the Nineties*. New York: Crown, 1991.

Cooper, Kenneth H. *The New Aerobics for Women*. New York: Bantam Books, 1988.

———. *Preventing Osteoporosis: The Kenneth Cooper Method*. New York: Bantam Books, 1989.

———. *Aerobics*. New York: Bantam Books, 1990.

Cutler, Winnifred B., Ph.D. *Hysterectomy: Before and After*. New York: Harper & Row, 1988.

Doress, Paula Brown, and Diana Laskin Siegal. *Ourselves, Growing Older*. New York: Simon and Schuster, 1987.

Friedan, Betty. *The Feminine Mystique*. New York: W. W. Norton, 1983.

Frisch, Melvin, M.D. *Stay Cool Through Menopause*. Los Angeles: The Body Press, 1989.

Gillespie, Clark, M.D. *Hormones, Hot Flashes and Mood Swings*. New York: Harper & Row, 1989.

Hausman, Patricia, and Judith Benn Hurley. *The Healing Foods*. Emmaus, PA: Rodale Press, 1989.

Henig, Robin Marantz, and the Editors of *Esquire*. *How a Woman Ages*. New York: Ballantine/Esquire, 1985.

Katahn, Martin, Ph.D. *The T-Factor Diet*. New York: Bantam Books, 1989.

Lark, Susan M., M.D. *The Menopause Self Help Book*. Berkeley, CA: Celestial Arts, 1990.

Love, Susan, M.D., with Karen Lindsey. *Dr. Susan Love's Breast Book*. New York: Addison-Wesley, 1991.

McCleary, Kathleen. "The No-Gimmick Weight-Loss Plan." *In Health* magazine, December/January 1992, pp. 82–83.

Madaras, Lynda, and Jane Patterson, M.D., with Peter Schick, M.D. *Womancare: A Gynecological Guide to Your Body*. New York: Avon Books, 1981.

Morgan, Dr. Brian L. G., and Roberta Morgan. *Hormones: How They Affect Behavior, Metabolism, Growth, Development and Relationships*. Los Angeles: The Body Press, 1989.

Nachtigall, Lila, M.D., and Joan Rattner Heilman. *Estrogen: The Facts Can Change Your Life*. New York: HarperCollins, 1991.

Natow, Annette B., Ph.D., R.D., and Jo-Ann Heslin, M.A., R.D. *The Fat Attack Plan*. New York: Pocket Books, 1990.

Norfolk, Donald. *The Stress Factor*. New York: Simon and Schuster, 1977.

Rinzler, Carol Ann. *Feed a Cold, Starve a Fever: A Dictionary of Medical Folklore*. New York: Facts on File, 1991.

Rubenstein, Carin, Ph.D. "The American Health Sex Survey." *American Health*, December 1991, pp. 56–57.

Shangold, Mona, M.D., and Gabe Mirkin, M.D. *The Complete Sports Medicine Book for Women*. New York: Fireside/Simon and Schuster, 1985.

Sheehy, Gail. *Passages*. New York: Bantam Books, 1974.

Shephard, Bruce D., M.D., and Carroll A. Shephard, R.N., Ph.D. *The Complete Guide to Women's Health*. New York: Plume/Penguin, sec. rev. ed. 1990.

Shimer, Porter with Sharon Ferguson. *Prevention* magazine. "Unwind and Destress." July 1990, pp. 75–92.

Shock, Nathan W., Richard C. Greulich, Reubin Andres, et al. *Normal Human Aging: The Baltimore Longitudinal Study of Aging*. U.S. Dept. of Health and Human Services, 1984.

Tapley, Donald F., M.D., Thomas Q. Morris, M.D., Lewis P. Rowland, M.D., et al. *The Columbia University College of Physicians and Surgeons Complete Home Medical Guide*. New York: Crown, 1989.

University of California, Berkeley. *The Wellness Encyclopedia*. Boston: Houghton Mifflin, 1991.

U.S. Dept. of Health and Human Services, Public Health Service, and National Institutes of Health. *The Menopause Time of Life*. July 1986. NIH Publication No. 86-2461. (For copies, write to : NIA Information Center, 2209 Distribution Circle, Silver Spring, MD, 20910.)

U.S. Dept. of Health and Human Services, Public Health Service, and National Institutes of Health. *Older and Wiser— the Baltimore Longitudinal Study of Aging.* September 1989. NIH Publication No. 89-2797.

Utian, Wulf H., M.D., Ph.D., and Ruth S. Jacobowitz. *Managing Your Menopause.* New York: Fireside/Simon and Schuster, 1990.

Wilen, Joan, and Lydia Wilen. *Chicken Soup and Other Folk Remedies.* New York: Fawcett Columbine, 1984.

Appendix B

Other Professional Sources

Barrett-Connor, Elizabeth, M.D. "Postmenopausal Estrogen Replacement and Breast Cancer." *The New England Journal of Medicine:* August 3, 1989: 319–320.

Bergkvist, Leif, M.D., Ph.D.; Hans-Olov Adami, M.D., Ph.D.; Ingemar Persson, M.D., Ph.D.; et al. "The Risk of Breast Cancer After Estrogen and Estrogen-Progestin Replacement." *The New England Journal of Medicine*, August 3, 1989: Vol. 321, No. 5, pp. 293–297.

Colditz, Graham A., M.B., B.S.; Meir J. Stampfer, M.D.; Walter C. Willett, M.D.; et al. "Prospective Study of Estrogen Replacement Therapy and Risk of Breast Cancer in Menopausal Women." *Journal of the American Medical Association*, November 28, 1990: Vol. 264, pp. 2648–2653.

Elsevier Scientific Publishers Ireland Ltd. "Special Issue: 25 Years of Hormone Replacement Therapy." *Maturitas*, September 1990; Vol. 12, No. 3, pp. 159–319.

Goldman, Lee, M.D.; and Anna N.A. Tosteson, Sc.D. "Uncertainty About Postmenopausal Estrogen." *The New England Journal of Medicine*, 1991; Vol. 325, pp. 800–802.

Grisso, Jeane Ann, M.D., M.Sc.; Jennifer L. Kelsey, Ph.D.; Brian L. Strom, M.D., M.P.H.; et al. "Risk Factors for Falls as a Cause of Hip Fracture in Women." *The New England Journal of Medicine*, 1991; Vol. 324, pp. 1326–1331.

Lobo, Rogerio A., M.D. "Cardiovascular Implications of Estrogen Replacement Therapy." *Obstetrics and Gynecology*, April 1990; Vol. 75, No. 4, pp. 18S–25S.

Matthews, Karen A.; Rena R. Wing; Lewis H. Kuller; et al. "Influences of Natural Menopause on Psychological Characteristics and Symptoms of Middle-Aged Healthy Women." *Journal of Consulting and Clinical Psychology*, 1990. Vol. 58, No. 3, pp. 345–351.

Menopause Digest, March 1991.

Menopause Management. Vol. IV, No. 1, 1991.

National Institute on Aging. *Special Report on Aging*, 1991.

Padwick, M. L., M.B.; J. Endacott, S.R.N.; and M. I. Whitehead, M.B. "Efficacy, Acceptability, and Metabolic Effects of Transdermal Estradiol in the Management of Postmenopausal Women." *American Journal of Obstetrics and Gynecology*, August 15, 1985; Vol. 152, No. 8, pp. 1085–1091.

Pines, Amos, M.D.; Enrique Z. Fisman, M.D.; Yoram Levo, M.D.; et al. "The Effects of Hormone Replacement Therapy in Normal Postmenopausal Women: Measurements of Doppler-Derived Parameters of Aortic Flow." *American Journal of Obstetrics and Gynecology*, March 1991; Vol. 164, No. 3, pp. 806–912.

Postgraduate Medicine: A Special Report. "Hormone Replacement Therapy—Where We Stand Now." New York: McGraw-Hill, 1990.

Sarrel, Philip M., M.D. "Sexuality and Menopause." Supplement to *Obstetrics and Gynecology*, April 1990; Vol. 75, No. 4.

Sarrel, Philip M., M.D. "Sexuality in Menopause." *Menopause Management*, Winter 1989, pp. 9–10.

Sitruk-Ware, Regine, M.D., and Wulf H. Utian, M.D., Ph.D. *The Menopause and Hormonal Replacement Therapy*. New York: Marcel Dekker, 1991.

Stampfer, Meir J., M.D.; Graham A. Colditz, M.B., B.S.; Walter C. Willett, M.D.; et al. "Postmenopausal Estrogen Therapy and Cardiovascular Disease—Ten-Year Follow-up from the Nurses' Health Study." *The New England Journal of Medicine*, 1991; 325: pp. 756–62.

Stanczyk, Frank Z., Ph.D.; Donna Shoupe, M.D.; Victoria Nunez, L.V.N.; et al. "A Randomized Comparison of Nonoral Estradiol Delivery in Postmenopausal Women." *American Journal of Obstetrics and Gynecology*, 1988; Vol. 159, pp. 1540–1546.

Walsh, Brian W., M.D.; Isaac Schiff, M.D.; Bernard Rosner, Ph.D.; et al. "Effects of Postmenopausal Estrogen Replacement on the Concentrations and Metabolism of Plasma Lipoproteins." *The New England Journal of Medicine*, October 24, 1991; Vol. 325, No. 17, pp. 1196–1204.

Whitehead, Malcolm I., M.B., B.S., F.R.C.O.G.; and David Fraser, M.B., B.S., B.Sc. "Controversies Concerning the Safety of Estrogen Replacement Therapy." *American Journal of Obstetrics and Gynecology*, May 1987; Vol. 156, No. 5, pp. 1313–1322.

Whitehead, Malcolm I., M.B.; M. L. Padwick, M.B.; J. Endacott, S.R.N.; et al. "Endometrial Responses to Transdermal Estradiol in Postmenopausal Women." *American Journal of Obstetrics and Gynecology*, August 15, 1985; Vol. 152, No. 8, pp. 1079–1084.

Whitehead, Malcolm I., M.B., B.S., F.R.C.O.G.; T. C. Hillard, B.M., M.R.C.O.G.; and D. Crook, Ph.D. "The Role and Use of Progestogens." *Obstetrics and Gynecology*, April 1990; Vol. 75, No. 4, pp. 59S–76S.

Appendix C

Papers Presented

Bachmann, Gloria, M.D. "Sexuality: The Role of Declining Gonadal Hormones." The North American Menopause Society 2nd Annual Meeting, September 25–28, 1991; Montreal, Canada.

Barrett-Connor, Elizabeth, M.D. "Estrogen and Cancer." The North American Menopause Society 2nd Annual Meeting, September 25–28, 1991; Montreal, Canada.

Pike, Malcolm, M.D. "Achieving Progestin Benefits at Minimal Risks." The North American Menopause Society 2nd Annual Meeting, September 25–28, 1991; Montreal, Canada.

Sarrel, Philip, M.D. "Sex and Menopause." The North American Menopause Society 2nd Annual Meeting, September 25–28, 1991; Montreal, Canada.

Steinberg, Karen, Ph.D. "Estrogen Replacement Therapy and Breast Cancer—A Meta-Analysis." The North American Menopause Society 2nd Annual Meeting, September 25–28, 1991; Montreal, Canada.

Appendix D

Self-Help Resources

Action on Smoking and Health (ASH)
2013 H Street, N.W.
Washington, DC 20006

American Association of Retired Persons (AARP)
1909 K Street, N.W.
Washington, DC 20049

American Cancer Society
1599 Clifton Road
Atlanta, GA 30329

American Dental Association
Department of Public Information and Education
211 East Chicago Avenue
Chicago, IL 60611

American Diabetes Association
National Service Center
1660 Duke Street
Alexandria, VA 22314
Call toll-free: 1-800-232-3472

American Dietetic Association
430 North Michigan Avenue
Chicago, IL 60611

American Foundation for the Blind
15 West 16th Street
New York, NY 10011

American Heart Association
7320 Greenville Avenue
Dallas, TX 75231

American Lung Association
1740 Broadway, P.O. Box 596
New York, NY 10019

American Society of Plastic
and Reconstructive Surgeons
233 North Michigan Avenue, Suite 1900
Chicago, IL 60601

Cancer Information Service
(a program of the National Cancer Institute)
Call toll-free: 1-800-4-CANCER
Or write:
National Cancer Institute
9000 Rockville Pike
Building 31, Room 10A24
Bethesda, MD 20892

Elderhostel
(educational experiences for older adults
which are based on campuses throughout the country)
100 Boylston Street, Suite 200
Boston, MA 02116

Help for Incontinent People
P.O. Box 544A
Union, SC 29379

HERS Foundation
Hysterectomy Educational Resources and Services
422 Bryn Mawr Avenue
Bala Cynwyd, PA 19004

National Council on the Aging (NCOA)
600 Maryland Avenue S.W.
West Wing 100
Washington, DC 20024

National Council on Alcoholism, Inc.
1151 K Street N.W., Suite 320
Washington, DC 20005

National Eye Care Project
American Academy of Ophthalmology
P.O. Box 6988
San Francisco, CA 94120
Call toll-free: 1-800-222-EYES

National Institute on Aging (NIA)
Information Center
P. O. Box 8057
Gaithersburg, MD 20898

National Institute of Arthritis
and Musculoskeletal and Skin Diseases
NIAMS Clearinghouse
Box AMS
Bethesda, MD 20892

National Institutes of Health
Federal Building, Room 6C12
Bethesda, MD 20892

National Kidney and Urologic Diseases
Information Clearinghouse
P. O. Box NKUDIC
Bethesda, MD 20892

The National Osteoporosis Foundation
2100 M Street N.W.
Suite 602, Dept. V.F.
Washington, DC 20037

National Society to Prevent Blindness
500 East Remington Road
Schaumburg, IL 60173

North American Menopause Society
c/o University Hospitals
Department of Ob/Gyn
2074 Abington Road
Cleveland, OH 44106

Office on Smoking and Health
5600 Fishers Lane
Park Bldg., Room 1-10
Rockville, MD 20857

Sex Information and Education Council
of United States
80 Fifth Avenue, Suite 801
New York, NY 10011

The Skin Cancer Foundation
245 Fifth Avenue
Suite 2402
New York, NY 10016

Index

About the Author

Ruth S. Jacobowitz is an award-winning medical writer and a former vice president at Mt. Sinai Medical Center, a large Cleveland teaching hospital, and she has headed her own medical public relations firm. Her many lectures on menopause have taken her all over the country and she has been featured on such television shows as the *CBS Morning Show* and *Sonya Live*, as well as in major newspapers and magazines. She is a founding member of the North American Menopause Society and a former Midwest chair of the Association of American Colleges Group on Public Affairs. She is the mother of three married daughters and grandmother of four. She and her husband, Paul, live in Cleveland, Ohio.